Materials, Matter & Particles
A Brief History

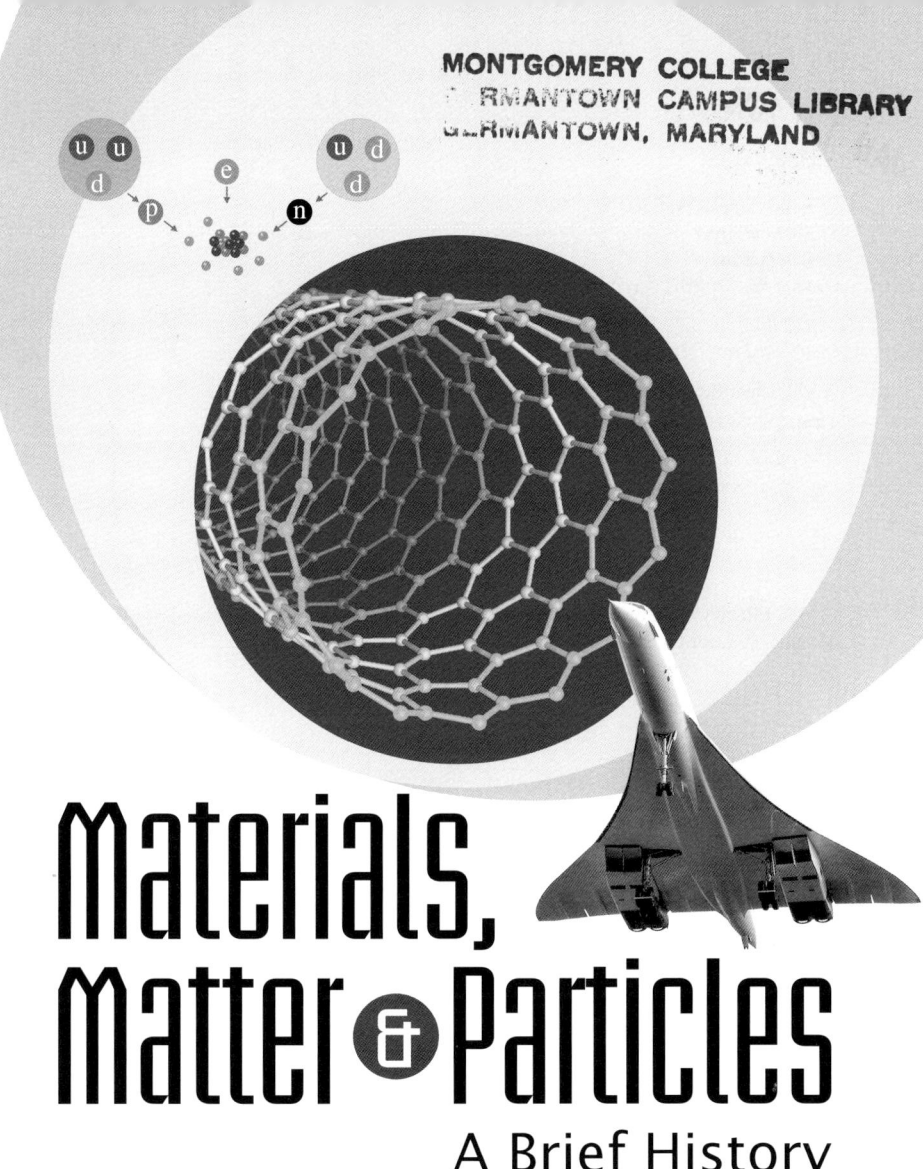

MONTGOMERY COLLEGE
GERMANTOWN CAMPUS LIBRARY
GERMANTOWN, MARYLAND

Materials, Matter & Particles
A Brief History

Michael M Woolfson
University of York, UK

Imperial College Press

Published by

Imperial College Press
57 Shelton Street
Covent Garden
London WC2H 9HE

Distributed by

World Scientific Publishing Co. Pte. Ltd.
5 Toh Tuck Link, Singapore 596224
USA office: 27 Warren Street, Suite 401-402, Hackensack, NJ 07601
UK office: 57 Shelton Street, Covent Garden, London WC2H 9HE

British Library Cataloguing-in-Publication Data
A catalogue record for this book is available from the British Library.

MATERIALS, MATTER AND PARTICLES
A Brief History

Copyright © 2010 by Imperial College Press

All rights reserved. This book, or parts thereof, may not be reproduced in any form or by any means, electronic or mechanical, including photocopying, recording or any information storage and retrieval system now known or to be invented, without written permission from the Publisher.

For photocopying of material in this volume, please pay a copying fee through the Copyright Clearance Center, Inc., 222 Rosewood Drive, Danvers, MA 01923, USA. In this case permission to photocopy is not required from the publisher.

ISBN-13 978-1-84816-459-8
ISBN-10 1-84816-459-9
ISBN-13 978-1-84816-460-4 (pbk)
ISBN-10 1-84816-460-2 (pbk)

Typeset by Stallion Press
Email: enquiries@stallionpress.com

Printed in Singapore by Mainland Press Pte Ltd

Contents

Introduction		1
Chapter 1	The Elements of Nature	5
Chapter 2	Early Ideas of the Nature of Matter	11
Chapter 3	The Quest for Gold and Eternal Life	17
Chapter 4	The Beginning of Chemistry	23
	4.1 The Chaos of Alchemy	23
	4.2 Paracelsus and His Medicines	24
	4.3 Robert Boyle, the Gentleman Scientist	25
Chapter 5	Modern Chemistry is Born	31
	5.1 The Phlogiston Theory	31
	5.2 Joseph Priestley	32
	5.3 Antoine Lavoisier	36
Chapter 6	Nineteenth Century Chemistry	41
	6.1 Chemistry Becomes Quantitative	41
	6.2 John Dalton	42
	6.3 Amedeo Avogadro	45

	6.4	The Concept of Valency	47
	6.5	Chemical Industry is Born	48
	6.6	Bringing Order to the Elements	50
Chapter 7	Atoms Have Structure		53
	7.1	Michael Faraday	53
	7.2	The Nature of Cathode Rays	55
	7.3	J. J. Thomson and the Electron	58
	7.4	The Charge and Mass of the Electron	61
Chapter 8	Radioactivity and the Plum-Pudding Model		65
	8.1	Röntgen and X-rays	65
	8.2	Becquerel and Emanations from Uranium	68
	8.3	The Curies and Radioactivity	70
	8.4	Rutherford and the Nuclear Atom	71
Chapter 9	Some Early 20th Century Physics		77
	9.1	The Birth of Quantum Physics	77
	9.2	The Photoelectric Effect	80
	9.3	Characteristic X-rays	82
Chapter 10	What is a Nucleus Made Of?		85
	10.1	First Ideas on the Nature of the Nucleus	85
	10.2	Moseley's Contribution to Understanding Atomic Numbers	86
	10.3	Breaking Up the Nucleus	89
	10.4	Another Particle is Found	91
Chapter 11	Electrons in Atoms		95
	11.1	The Bohr Atom	95
	11.2	Waves and Particles	99
	11.3	The Bohr Theory and Waves	101
	11.4	An Improvement of the Bohr Theory	103

Chapter 12	The New Mechanics		107
	12.1	Schrödinger's Wave Equation	107
	12.2	The Wave Equation and Intuition	109
	12.3	Orbits Become Orbitals	112
	12.4	The Great Escape	116
	12.5	Heisenberg and Uncertainty	118
Chapter 13	Electrons and Chemistry		121
	13.1	Shells and the Periodic Table	121
	13.2	Valency	125
Chapter 14	Electron Spin and the Exclusion Principle		129
	14.1	The Stern–Gerlach Experiment	129
	14.2	Electrons in a Spin	131
	14.3	The Pauli Exclusion Principle	133
Chapter 15	Isotopes		135
	15.1	What is an Isotope?	135
	15.2	The Stable Isotopes of Some Common Materials	136
Chapter 16	Radioactivity and More Particles		141
	16.1	The Emission of α Particles	141
	16.2	β Emission	143
	16.3	The Positron	144
	16.4	Another Kind of β Emitter	147
	16.5	The Elusive Particle — The Neutrino	148
	16.6	How Old is that Ancient Artefact?	152
	16.7	An Interim Assessment of the Nature of Matter	153
Chapter 17	Making Atoms, Explosions and Power		155
	17.1	The New Alchemy	155
	17.2	Reactions with α Particles	155

17.3	Reactions with Protons	157
17.4	γ-ray Induced Reactions	157
17.5	Neutron Induced Reactions	158
17.6	Fission Reactions	159
17.7	The Atomic Bomb	162
17.8	Atomic Power	165
17.9	Fusion — Better Power Production and Bigger Bombs	166

Chapter 18 Observing Matter on a Small Scale — 169

18.1	Seeing	169
18.2	Microscopes	171
18.3	X-ray Diffraction from Crystals	176

Chapter 19 Living Matter — 183

19.1	Defining Life	183
19.2	Forms of Life	187

Chapter 20 Life at the Atomic Level — 195

20.1	Seeing Life Matter at the Atomic Level	195
20.2	Encoding Complex Structures	197
20.3	Encoding Living Matter — The Chemistry of DNA	198
20.4	The Double Helix	203
20.5	What Makes Us How We Are?	207
20.6	The Artificial Manipulation of Genes	210

Chapter 21 Materials from Ancient Times — 213

21.1	The Earliest Use of Natural Materials — The Stone Age	213
21.2	Some Early Manufactured Materials — The Bronze Age	215
21.3	The Iron Age	222
21.4	Cement and Concrete	227
21.5	Clothing Materials	230

Chapter 22	Modern Materials		237
	22.1	Materials Old and New	237
	22.2	Manufactured Fibres, Synthetic Fibres and Plastics	238
	22.3	Semiconductors	245
	22.4	Nanotechnology	250
Chapter 23	The Fantastic World of Particles		255
	23.1	Antics on Ice	255
	23.2	The First Designed Atom-Smasher	259
	23.3	The Cyclotron	261
	23.4	The Stanford Linear Accelerator	262
	23.5	Synchrotrons and the Large Hadron Collider	263
	23.6	Some Fundamental Particles	267
Chapter 24	How Matter Began		273
	24.1	Wailing Sirens	273
	24.2	Measuring the Universe	275
	24.3	The Expanding Universe	278
	24.4	The Big Bang Hypothesis	280
	24.5	The Creation of Matter	281
Chapter 25	Making Heavier Elements		287
	25.1	The Architecture of the Universe	287
	25.2	Dark Matter and Dark Energy	290
	25.3	The Formation of a Star	291
	25.4	The Death of a Star	294
	25.5	The Evolving Universe	299
Index			305

Introduction

Civilized societies are characterized by the widespread use of a range of materials to construct the artefacts on which that society depends. However, within any such society there will always be an intelligentsia of some sort — priests, teachers and philosophers and so on — who will ponder about the nature of matter, the stuff that constitutes our world, and the individual materials that are the manifestation of that matter. The first ideas about matter in some of the more sophisticated earlier civilizations, such as that of ancient Greece, is that there were just a few simple substances from which all material was constituted. The Greeks, and others, chose the substances air, earth, fire and water as a basic set having a range of properties that would be possessed by all other material. If a material, say a metal, had rigidity then a major part of it would have to be earth, since the other three substances had no rigidity whatsoever. Again, something like honey would have a large component of water since it flows like water but, unlike air and fire, it maintains a fixed volume within a container.

One of the earliest drivers that stimulated the systematic investigation of matter was alchemy, the desire to make gold from base metals and to find the means of extending life — to eternity if possible. Although we now know that these activities were futile in terms of achieving their desired ends, they did nevertheless achieve something quite important. From the ranks of the alchemists there arose a class of individuals now called scientists, the seekers after knowledge for its own sake, and they created the first science — chemistry. From the 17th century onwards knowledge about the nature of matter grew apace and chemists gradually built up an inventory of elements, the atoms that are the constituents of all materials. At first there seemed to be very little relationship between the individual elements they discovered but then a pattern of relationships emerged, connecting elements with similar properties, connections which came to be represented in a tabular form — the Periodic Table.

This neat chemists' world of a fairly large, but finite, number of elements with indivisible atoms started to crumble at the end of the 19th century when physicists began to explore the nature of atoms. This was not actually the goal that they were initially pursuing. The physicists were interested in the way that electricity was conducted through gasses at very low pressure and the phenomenon of radioactivity. Starting with this seemingly unrelated work, step by step they built up a picture of an atom, not as something indivisible but as something with structure that could break down, either spontaneously or by being bombarded in some way. All atoms were now seen as structures consisting of protons, neutrons and electrons, so seeming — with only three components — to return to the simplicity of the early Greeks.

However, having apparently reintroduced straightforwardness into the description of the material world, the physicists, with further work, soon disturbed that simple picture. Isaac Newton (1642–1727) had described the behaviour of material bodies in terms of the laws of mechanics that accurately described the behaviour of bodies as disparate as billiard balls and spacecraft. It turned out that Newton's laws were inadequate for very tiny particles such as electrons; a new mechanics, wave mechanics, was discovered that could deal with the behaviour of such tiny particles. In this new mechanics the particles are described as though they have wave-like properties, similar to that of light, and there arose a new concept — wave-particle duality — where, sometimes, material particles could behave like waves and where radiation, usually described in terms of waves, could occasionally behave like particles. To add to this complication, in their high-energy collision experiments physicists discovered large numbers of short-lived exotic particles that seem to have little relevance to the material world in which we live. Fortunately it turned out that not all these exotic particles were basic and many of them, together with protons and neutrons, were composed of even more fundamental particles — quarks. It is just possible that we may not yet have reached the end of the story about what constitutes the fundamental particles of matter!

From the very earliest times humankind has exploited materials in a great variety of ways. This includes naturally occurring materials such as stone or wood but quite early it was found that metals could be extracted from particular rocks and that some mixtures of metals had better properties than the component metals themselves. In more recent times great industries have been built on the creation of new materials — textiles, plastics, pharmaceuticals and semiconductors, for example. New knowledge concerning the chemical basis of life and living materials — the structures of deoxyribonucleic acid (DNA) and proteins — have given prospects of great medical advances through gene therapy and possible benefits from genetically modified plants, either for food or for other uses.

The final topic of this book is the most fundamental of all and that is how all this material came into existence in the first place. That goes back to the Big Bang, the beginning of time, of space and of all the material in the Universe, including the infinitesimal part of it that constitutes our world.

Chapter 1

The Elements of Nature

Despite the canopy that shielded the corner of the courtyard that served as the household school, the sunny, dry and windless day was oppressive in the extreme. Acacius, the educated slave procured by Rufus for the express purpose of teaching his sons, Marcus and Tacitus, estimated from the position of the Sun that there were two hours still to pass before he would be free of his troublesome charges, at least for the duration of the midday break. It was not that the boys weren't bright — on the contrary they were both very clever — but on this aggressively hot day they would much sooner be fishing in the river within the shade of the nearby forest. And so would Acacius for that matter. He had planned to concentrate on Greek grammar during this morning session of the school but it was not a subject calculated to maintain the boys' interest on such a day. It would be better to find another subject, one where he could exercise the Socratic principle of teaching in which knowledge was created by the students themselves in answering questions put to them by their teacher. In that way he could at least hope to keep their minds working and their attention engaged.

First he must bring their minds towards the topic of his alternative lesson. 'Boys, today we are going to explore what it is that constitutes our world. We look around us and we see that it is very complicated and that it consists of many different entities. Some of these are inanimate and unchanging like the stones on the mountainside, others do not live but they do change, like the clouds in the sky. Then there are the living things in our world — things that are born, grow old and die — like goats, trees and even men. The philosophers of old considered the nature of these constituents of our world and concluded that they were all different combinations of five different principles, or elements, four of them material and the fifth belonging to the spiritual world. The material elements are inanimate and owe nothing to man's influence. We shall first consider these.'

Acacius decided to put the first question to the older boy, Marcus. Tacitus, as his name indicated, was not a very forthcoming individual and his answers were inclined to be monosyllabic and terse in the extreme. Marcus, by contrast, could be over-elaborate in his answers to questions and would frequently wander into areas not directly related to the original question. It was surprising that two boys of such different characters, separated by just over a year in age, could both have come from the union of Rufus and Justina. 'Marcus, look around you in every direction. What do you see that comes from nature itself and owes nothing to man's work and invention?' Marcus looked around. He could see the house — but that was man-made — as was the table with the pitcher of water and the beakers that had been put there for their benefit. What a difficult question — everything around was man-made. He spent several minutes pondering the question — Acacius was insistent that one should not hurry the process of thinking through problems — but Marcus's mind was blank. He idly brushed some dust off his hand that had been resting on the ground and then inspiration came. 'Dust' he exclaimed 'All around is dust and that is not man-made. The farmers in the fields...' but at that point Acacius cut him short. Given his head Marcus would launch into a discourse on farming and eventually branch out into an explanation of the shortage of imports of grain from Africa, relaying what he had overheard in a conversation between his father and his uncle Drusus. 'An excellent answer Marcus but perhaps we could use another word instead of dust. That farmer you were going to tell me about — does he till dust?' 'No, of course not, he tills the earth' said Marcus, and then the light dawned. 'That's what my answer is — earth.'

Acacius felt he was making progress — the Socratic method was yielding results and already one of the elements of nature had come from the lips of one of his pupils. Now Acacius must try to elicit some information from the other brother. He inwardly blanched. Although Tacitus was actually the brighter boy of the two, getting information out of him was like trying to squeeze oil out of an olive after the olive had already been through the press — possible but very difficult. 'Now Tacitus what more do we have around us that is part of nature and is not made by man?' Tacitus looked around him and his eyes alighted on the pitcher. That was certainly made by man but what was inside it was not. It contained water and that came by the blessing of Jupiter, who was not only the king and master of all the gods but who controlled the sky, the clouds in the sky and the rain that came from those clouds. All these thoughts were concentrated in one word — 'Water' Tacitus grunted and then fell into a deep silence. 'Yes Tacitus, but wouldn't you like to expand on that a little' Acacius weakly protested. Tacitus grudgingly obliged 'Element of nature — water' he replied, forcing himself into unfamiliar verbosity to please his tutor. He liked Acacius and he was a scholarly boy who enjoyed learning new things but he did not see why one had to be garrulous in the process. Acacius appreciated that Tacitus had made a supreme effort on his behalf and decided to move on.

'Now back to you, Marcus. The next element I want us to find is a difficult one because it is invisible — but it is everywhere. Think really hard about what it might be.' Marcus was bewildered looking left and right, up and down, although, since what he sought was invisible he realised that he might just as effectively think with his eyes shut. After several minutes he gave up — how could something be everywhere but invisible? Acacius thought hard; he felt that to tell the boys directly what the answer was would be a betrayal of the Socratic method so he had to find a way for the answer to come from Marcus' own lips. Lips — that might be the answer. He gave instructions to Marcus — 'I want you to close your mouth and push your lips together hard. Now pinch your nose hard between your forefinger and thumb until I tell you to stop doing so.' Marcus did as he was told but within a minute he was going red in the face, opened his mouth and gasped for breath. 'You disobeyed my instructions' Acacius said to him sternly. Marcus protested 'I couldn't breathe — I thought I was going to die.' 'But why did you need to breathe'

Acacius asked him. Marcus began to respond 'All living things need to ...' and then a look of delight suffused his face and excitedly he exclaimed 'I know, I know — it is air, it is air.'

Acacius was triumphant; he had succeeded in getting the answer he wanted. 'Well done Marcus, but now we must find our final, and perhaps hardest, element of nature. That will be your task Tacitus.' Tacitus, as usual, showed no emotion. He usually succeeded in doing what he had to do and this time would be no different. 'Your task, Tacitus, is to find an element of nature that can be seen but that would be sensed by a man at a distance even if he was blind.' Tacitus thought hard. What is it that a blind man could sense? It couldn't be smell because that can't be seen. It couldn't be touch because that can't really be seen — only the thing being touched and that could be anything. If only it wasn't such a hot day he could think more clearly. That is a possibility; a blind man could detect heat but heat as such cannot be seen. What *can* be seen is a fire that produces the heat and that must be the answer. He had better give a longer answer this time to keep Acacius happy — 'The fourth element of nature is fire' Tacitus stated. Acacius was delighted and amazed. He had thought it would require quite an effort on his part to tease the answer from Tacitus but the boy was very clever, exceptionally so. 'That is splendid Tacitus. We now know the four elements of nature in terms of which all the material contents of the world are constituted. Now I am going to write them down so we can see them all together.' Acacius picked up a stylus and, on his wax tablet, he inscribed the following:

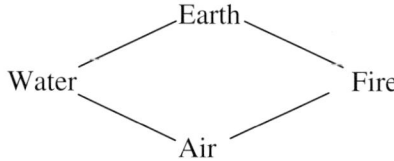

For the next part of his lesson Acacius did not think that he would make much progress by the Socratic method. The Greek philosophers of yesteryear had come up with qualities that linked the elements in pairs but in his heart of hearts Acacius was not fully convinced by the logic of some of their choices. It would be better just to quote these connections and brook no arguments — once the wisdom of the past was challenged that would be the beginning of the end of civilization. 'Now boys, the philosophers of old, in their infinite wisdom, have seen how these elements are connected to each other.' Acacius then modified the diagram on the wax tablet as follows:

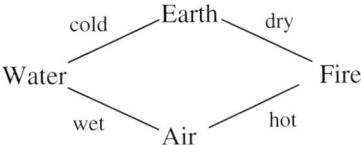

Acacius then explained. 'Although the elements can have various qualities — for example, water can be cold or hot — more often than not the following qualities can be associated with each of the elements

>Earth is cold and dry
>Fire is dry and hot
>Air is hot and wet
>Water is wet and cold.'

Tacitus said nothing but was dubious. The earth he was sitting on was certainly not cold, although it was true it was very dry. He also remembered a period he had spent in the mountains during the previous winter when the air had been distinctly cold. Perhaps the world had been different when the old philosophers had lived and worked. Marcus also said nothing — after all, if Acacius said that something was true then it was so.

Now Acacius moved on to the conclusion of the lesson. 'The elements that you have discovered this morning are, indeed, the basic ones for describing the characteristics of the inanimate material objects of our world. Even the air, which is invisible to us, is a material thing, as we know when Favonius[a] exerts his full strength. But

---

[a] Favonius was the Roman god of the west wind.

there is something beyond the material world, some element that would not be a constituent of a rock or a mountain stream. Can you think of any such thing?' The boys thought hard with wrinkled brows. Acacius gave them a clue. 'Your brows are wrinkled. Why is that?' 'Because we are thinking' replied Marcus and, almost immediately afterwards, Tacitus said 'Rocks and mountain streams do not think — the other element is thought.' He was so excited by this idea that he was stimulated into extreme verbosity. Acacius beamed at him — 'Yes, you are right, but perhaps a better word is "Idea" the product of thought. Now we have all the elements — four of them material and one ethereal that characterize everything that exists. All the material objects in the world can be described in terms of the extent to which they possess the qualities of the four material elements. The products of the mind — for example, ideas, imagination and the dreams that accompany our sleep — they come within the realm of the fifth element, known as "Idea" or "Quintessence". While you are having your midday break I would like you to think of four different entities and the way in which they may be described in terms of the basic elements. We shall discuss your ideas in afternoon school.'

Acacius noted with relief the approach of the slave announcing the time for the midday meal. Three precious hours when he could engage with his own thoughts — after which, back to the boys. Perhaps later in the day, when it was cooler and after discussing the boys' contributions about the elements, he would cover the grammar that he had planned for the morning session.

Chapter 2

Early Ideas of the Nature of Matter

The basis of western philosophy was founded in classical Greece in the sixth century BC. The early Greek philosophers were concerned with the nature of the world, both in a material and spiritual sense, and they believed that the structure of the world and all in it had to conform to concepts of perfection and harmony. For example, quite early on they postulated that stars and planets were separate material bodies and they argued that these bodies had to be spheres, based on the principle that the sphere is the shape of greatest 'perfection'. The sphere is the only shape that is seen in precisely the same way — as a circle — from any perspective, an idea in accord with the Greek concept of perfection and harmony.

Greek philosophers would have been uncomfortable with the idea that every distinctive material had an independent identity, unrelated to that of any other material, meaning that an inventory of all the types of distinctive material in the world would have been huge and virtually infinite. The Greeks regarded simplicity as a desirable characteristic of any theory, a view shared by modern scientists. Consequently, the theory that they accepted was that all the various kinds of material were merely different manifestations either of a single substance or of mixtures of just a few basic substances. Anaximenes, a sixth century BC Greek philosopher, took air as the single basic substance with everything else produced by concentrating or diffusing it to form other materials and fire. However, in the fifth century BC

Empedocles expanded the list to the four basic principles of air, earth, fire and water, the theory that Acacius explored with his pupils. The idea of the four basic elements, or principles, became well established and was accepted even well into the medieval era.

It is certain that, in their considerations of the nature of matter, the Greek philosophers were greatly influenced by ideas that originated in civilizations much older than their own. Greek civilization began about 2000 BC but at the time it arose there were already well-established civilized communities in Mesopotamia, Egypt, India, Persia and China — and by the time of Empedocles other civilized communities existed, for example, in Japan. Ideas about the nature of the material and spiritual world were prevalent in many early civilizations and were remarkably similar to each other. Although personal travel over the vast distances from one extreme region to another of the early civilizations, say from Egypt to Japan, was almost non-existent that is not to say that ideas could not permeate across those distances. The transport of ideas, given long periods of time for their passage over the boundaries of contiguous societies, was far more efficient than the transport of people. Thus the Indians and Japanese shared with the Greeks the material elements of air, earth, fire and water, although the Japanese substituted 'wind' for 'air'. However, the early civilizations varied somewhat in their descriptions of the spiritual element. To the Greeks the fifth element, the *quintessence*, was the stuff that the gods were made of while to the Japanese this element, the *void*, just represented the quality of anything not of the material world. By contrast, the Chinese, being a rather pragmatic and materialistic society, had five basic material elements — earth, fire, water, wood and metal — apparently without any spiritual element although they had something called *qi*, a form of energy of which heaven consisted.

At a local level there was a uniformity of ideas in the Mediterranean region, the source of most advanced thinking in the Western world up to medieval times. When Acacius, of Greek origin, taught his Roman pupils he was teaching them the philosophy shared by Romans and Greeks and most other peoples in that region. What

Acacius was teaching his pupils was just about the basic elements from which matter was formed rather than the precise properties of that matter. Of course the Greeks knew that there were distinct materials — stone, wood, clay, iron, bronze, fur, wool and so on — with different properties and they exploited them in their everyday lives. The important concept, at least to the intellectuals of Greek society, was that, although these various materials seemed rather different and had a great variety of properties and uses, they were all manifestations of the four basic elements from which they derived their properties.

An important concept in their understanding of the nature of materials was the role of fire. With fire one could transform wood into ash, something similar to earth, and that same ash, mixed with sand and subjected to fire, could be transformed into glass, which resembled none of its components. Similarly, by the action of fire, rocks could be made to yield metal and different metals could be combined to make new metals. Bronze, made by combining copper and tin, was a harder and stronger metal than either of the components and would be regarded as a distinctive metal made by a transmutation to a completely new material, rather than as being some combination of the two original components. Thus the material world contained a vast number of individual materials with different properties, and transformations, usually wrought by fire, could sometimes change one material, or combination of materials, into another. However, underpinning the whole system there were the four basic elements from which they were all constituted.

Two Greeks from the fifth century BC, Leucippus (Figure 2.1) and his pupil Democritus (Figure 2.2), put forward ideas about the nature of materials that have a resonance today. They considered the process of repeatedly dividing a material by cutting it into halves. They concluded that, eventually, one would arrive at an amount of material that could not be divided further. These indivisible units were designated by the Greek word '*atoma*' (indivisible objects) from which we get the term *atom*, the units of matter once thought by chemists to be indivisible. Democritus was, by all accounts, a curious man, thought by some to be mad on account of his habit of laughing

14   *Materials, Matter and Particles*

Figure 2.1   Leucippus.

Figure 2.2   Democritus.

in maniacal fashion while commenting on the behaviour of his fellow men — but probably he just had a good sense of humour and saw the funny side of life. However, mad or not, he came up with ideas that for their time were a great advance on what had preceded them. He argued that the physical properties of matter were related to the physical nature of the atoms of which they consisted. Thus a fluid like water consists of large spherical atoms that can easily move relative to each other and so give it fluidity. If the spherical atoms were small then the relative motion would be enhanced so that the liquid would have a slippery or oily nature. By contrast a solid would consist of small rather jagged atoms that would resist moving past each other and so give a solid its rigidity. For very rigid solids, such as metals, this idea was extended to having hooks on the atoms that bound them together, thus inhibiting any relative motion. These ideas, relating the physical attributes of a material to what was happening at the atomic level, are very much in tune with modern theories, albeit that the mechanisms proposed for the interactions are very different.

Ideas about basic indivisible units of matter were not confined to the Greeks. At about the time that Leucippus and Democritus were developing their model of basic units, based on a philosophical consideration of the nature of matter, the Indian sage Kanada, who founded the philosophical school known as Vaisheshika, was putting forward similar ideas, probably arrived at by meditation rather than by the Greek process of detailed logical analysis. Kanada believed in the four material elements of the Greeks but added a fifth, *Akasha*, which can be roughly translated as the *ether*, a substance that pervades the whole of space and is linked with the phenomenon of sound. Again, in the Vaisheshika philosophy there was not one non-material spiritual element, as in the Greek system, but four, which are *time, space, soul* and *mind*. The Indian philosophy included the concept of atomism but it was linked with the idea that the way that atoms behaved was governed by the influence of some Supreme Being — a curious mixture of materialism and theism.

The ideas we have described that originated in the ancient Greek world of the sixth century BC persisted up to about the 16th century and so had an acceptance life of 2 000 years or more — a remarkably long time for a philosophical, or scientific, idea to hold sway.

The Greeks, being of a philosophical bent, tended more towards abstract considerations of the nature of matter than to practical aspects. The fact that copper and tin, both soft metals, could be combined to give bronze, a useful hard and strong metal that could be used for making weapons or tools, suggested to them that materials could be transmuted from one form to another. However, there is no evidence that they followed up that example, of what would now be called 'material science', in any practical way. In other societies the practical approach played a much larger role.

Chapter 3

The Quest for Gold and Eternal Life

The citizens of one of the earliest civilizations, ancient Egypt, were obsessed with the idea of life after death. It was considered important for this future life that the body should be preserved in as intact a state as possible and over the course of time methods of mummification were developed that were extremely effective. The first preservation of bodies probably came about by accident rather than design. A body buried in sand in the very arid climate of Egypt would become desiccated and thus preserved, rather like biltong, the dried preserved meat that originated in South Africa. It might also have been found that a body would be even better preserved in a mixture of sand and salt: a combination of salting and drying is used to produce dried cod, notably in Norway, Iceland and Canada.

The Egyptian process of mummification was carried out by priests and was accompanied by a great deal of religious ritual. It was extremely complex and first involved removing softer tissues, such as the brain and internal organs, which would decay most rapidly. These were preserved in special containers, *canopic jars*, with each different type of organ in a distinctive jar bearing an image of an animal head that indicated the type of organ therein. Then the main part of the body was dried out for about forty days using natron, a natural product found in salt beds, which is a mixture of sodium carbonate decahydrate (soda ash), sodium bicarbonate (baking soda), sodium chloride (ordinary table salt) and sodium sulphate. Once the body

Figure 3.1 The golden mask from the tomb of Tutankhamun (Jon Bodsworth).

was thoroughly dried it was wrapped in bandages and placed in a sarcophagus, usually made of wood and ornately decorated with an image of the deceased (Figure 3.1).

In developing the processes of mummification the Egyptian priests were gaining knowledge of some basic chemistry. However, their goal was not limited to the protection of the body for a life after death. They sought also to extend life, even to the extent of finding a potion that would give eternal life. This was the tentative beginning of the practice that came to be known as *alchemy*. The term alchemy is thought to derive from the Greek word for Egypt 'Khemia', itself taken from the Egyptian term 'Khem' that referred to the dark soil deposited by the Nile in its annual floods. This word was adopted by the Greeks also to describe the Egyptian priestly practices for dealing with the dead and death. Later, in the seventh-century, the Arabs occupied Egypt and they added the Arabic prefix 'al' meaning 'the' and so the term alchemy came into use.

Some of the most important seminal developments in chemical understanding were due to the work of Muslim alchemists. The most famous of these was the eighth-century alchemist, Jabir ibn Hayyan (Figure 3.2), who is normally referred to by his Latinized name, Geber. He concluded that metals were intimate mixtures of the four elements — air, earth, fire and water — two of them as interior components and two as exterior components. According to his model

Figure 3.2 Jabir ibn Hayyan (721–815).

the rearrangement of these components could effect the transformation of one metal into another. Such changes required the use of a substance referred to in Arabic as *al iksir*, a term in use at the time to describe a potion consisting of some medically active ingredient dissolved in a solvent containing ethyl alcohol. This is the source of the modern word *elixir*, used to indicate a universal remedy for all ailments. It was thought to be a red powder derived from a stone — the *philosopher's stone* — and the search for this stone became the goal of alchemists throughout the ages. Geber thought that gold and silver were, in some way, hidden within base metals and by the right chemical processes that rearranged the elements within the base metals the precious metals could be produced. Although what he sought was impossible, Geber made significant advances in what is now the science of chemistry and he is sometimes given the accolade of having been the first chemist. His achievements were many. He recognised the difference between acid and alkaline materials and produced hydrochloric and nitric acids, a mixture of which, *aqua regia*, has the property of dissolving gold, which no other acid can do. He also developed the processes of distillation and crystallization, both essential in purifying materials.

The practice of alchemy was not confined to the western world — it also flourished in China and in India. In China the main practitioners were Taoist monks who were mainly concerned with the prolongation of life, in the pursuit of which they made significant advances in the development of herbal medicines, many of which were quite effective. These herbal medicines, and their successors, are the basis of Chinese and Indian medical practice today. A side-product of Indian alchemy was the production of high-quality steel where, by the use of bellows to create strong draughts of air, the right proportion of carbon was introduced into iron at a high temperature. Poorer quality steel had been produced earlier; the earliest steel, dating from about 1400 BC, originated in East Africa, not a region usually thought of as being in the vanguard of technology. Indian steel, knowledge of which was transmitted to the Arabs via Persia, was the probable source of 'Damascus steel' used from the eleventh-century for the making of swords of exceptional strength and sharpness — much to the discomfiture of the Crusaders.

The two goals of the alchemist, the transmutation of base metals into gold and the search for the 'elixir of life', a magical potion enabling those that took it to live forever in a state of eternal youth, were linked together through the concept of the philosopher's stone. Gold was seen as the perfect metal, incorruptible and unchanging with time. Any agency that could confer that property on a base metal was seen as potentially able to confer the same property to life, an attractive prospect at a period when human lifetime rarely extended beyond fifty years, and was often cut short by devastating plagues. However, the activities of alchemists did not meet with universal approval even in the mediaeval period since there was a suspicion that alchemists were engaged in trying to make 'false gold' and pass it off as the real stuff. In 1315 Pope John XXII, based in Avignon, banished all alchemists on the suspicion that they were engaged in counterfeiting and a 1403 law in England, passed by Henry IV, made the crime of 'multiplication of metals' punishable by death.

The idea that base metals could be turned into gold or that life could be extended indefinitely now seem so bizarre that we tend to think of them as the beliefs of a primitive society, or at least one well removed from contact with modern scientific thought. Yet the practice of alchemy carried on to comparatively recent times and was certainly a respected form of scientific activity in the medieval period (Figure 3.3).

The Quest for Gold and Eternal Life 21

Figure 3.3   The Alchemist (William Fettes Douglas, 1822–1891).

Figure 3.4   Isaac Newton (1642–1727).

As time moved on, and scientific endeavour was increasingly carried out in the name of enlightenment, so alchemy became less and less highly regarded, although some of its practitioners were very distinguished individuals. Isaac Newton (Figure 3.4), probably the greatest scientist who ever lived and who could be regarded as the grandfather of modern science, spent most of his later years in the practice of alchemy — although it was looked down on by the scientific establishment of the time. Despite the low regard in which alchemy was held, in retrospect it can be seen to have played an important role in the development of science. Although their experiments did not achieve their goals the alchemists were establishing a good foundation of practical knowledge about chemical reactions that eventually led to the important branch of modern science now known as chemistry.

During the fifteenth and sixteenth centuries, with gathering strength, the emphasis began to move away from the objectives of alchemy towards the study of chemistry as a scientific area of study, the purpose of which was to advance man's knowledge rather than to seek riches by making gold or to seek immortality by producing the elixir of life. By the seventeenth and eighteenth centuries, Newton notwithstanding, serious scientists were beginning to look down on alchemy as unworthy in its objectives and the subject of chemistry began in an earnest way.

Chapter 4

The Beginning of Chemistry

4.1 The Chaos of Alchemy

The practice of alchemy tended to be rather chaotic and alchemists on the whole were very secretive, not wishing to divulge their methods and discoveries, such as they were, to rival alchemists. Their descriptions of what they were doing were couched in terms that seemed designed to confuse rather than to inform the listener or reader. Very often they accompanied their coded accounts of what they had done with information about the phase of the Moon or the state of the tide at the time. While all this mystery may have given the impression that important activity was afoot, not to be understood by the ignorant *hoi polloi*, in fact most of the activities of alchemists were, for the most part, unsystematic and yielded very little information — even to the practitioners of alchemy themselves. By the end of the medieval period alchemists were not a highly regarded part of society and were suspected by many of those in power to be potential criminals, wishing to swindle society by the production of false gold and silver. However, by that time there was gradually evolving a tendency for chemical experiments to be carried out by people working in a more systematic fashion with the goal of understanding the rules of chemistry and for these people to record their findings in a clear and coherent way. The fool's gold of alchemy was transforming into the true gold of science!

4.2 Paracelsus and His Medicines

An early contributor to this passage from alchemy to chemistry was a Swiss doctor, Philippus Aureolus Paracelsus (Figure 4.1). The work and writings of Galen (129–200), a Greek doctor who became the court physician to the Roman emperor, Marcus Aurelius, dominated the practice of medieval medicine. Galen was noteworthy for his studies of human anatomy and for carrying out many surgical procedures, including some on the brain and eyes. His prime approach to medicine was dominated by surgery and, indeed, his advocacy of bloodletting as a treatment for many ills persisted until the 19th century. Galen had many followers in medieval Europe, and also in the Muslim world, but Paracelsus rejected his teachings. His alternative approach was the treatment of illness through medication and consequently his chemical studies were driven by medical priorities. He investigated many chemicals for their potential medical applications. Paracelsus was the first to use laudanum, a solution of opium in alcohol diluted with water, as a painkiller and he also discovered that mercury compounds could be effective in the treatment of syphilis. By the

Figure 4.1   Paracelsus (1493–1541).

15th century, Islamic scholars had moved away from the exclusivity of the four elements of air, earth, fire and water and added mercury and sulphur as elements that were the constituents of metals. Paracelsus, and some others with similar medical interests, accepted the idea of the four basic elements, as proposed by Greek philosophers. They also believed that there was an alternative description of substances that could be made in terms of the new elements mercury and sulphur with salt as a third component. In this trinity of substances mercury represented the property of flowing and volatility, sulphur the property of flammability and salt the property of inertness.

During the course of his investigations Paracelsus discovered many new substances, some organic, such as diethyl ether (now commonly known simply as *ether*, the volatile liquid that can be used as an anaesthetic), which he described as 'oil of sulphur'. He also divided substances into groups distinguished by their similar reactions to various processes, the kind of chemical classification made today, albeit in a different way. It was generally known that chemical remedies were much more effective if the materials were pure and that impurities could either reduce their effectiveness or give harmful side effects. To deal with this problem Paracelsus developed many new laboratory techniques for purifying substances including distillation and also freezing, which can, for example, concentrate alcohol in a water–alcohol mixture. This last process, now known as freeze distillation is (illegally) used for making the spirit Applejack from cider by freezing out some of the water it contains while retaining the alcohol.

4.3 Robert Boyle, the Gentleman Scientist

Another very important individual in this passage from alchemy to chemistry was the Irishman, Robert Boyle (Figure 4.2). He came from an extremely wealthy family, being the fourteenth child of Richard Boyle, the First Earl of Cork. His early education was at Eton College but at the age of eleven he left on an extended tour of Europe in the care of a French tutor. In 1641, at the age of 14, he spent a winter in Florence studying the work of the astronomer Galileo Galilei, who

Figure 4.2 Robert Boyle (1627–1691).

was living in that city under house arrest, sentenced by the Inquisition for supporting the theory that the Earth moved round the Sun. Robert Boyle returned to England in 1645, aged 18, and took up the life of a country gentleman with estates both in Ireland and in the county of Dorset, England, where he took up residence. His French tutor had done a good job in educating Robert Boyle for he acquired a curiosity about scientific matters that was to dominate the remainder of his life. He began to engage in scientific research and became a leading member of a group of like-minded individuals, known as the 'Invisible College' who were devoted to the development of science, which they referred to as the 'New Philosophy'. The group met on a regular basis, sometimes in Oxford and at other times at Gresham College in London. In 1654, to make more convenient his attendance at meetings of the Invisible College, Boyle moved to Oxford where he stayed until 1668 when, finally, he moved to London. From 1649, after the execution of Charles I, until 1660, the victors of the English Civil War, the Parliamentarians led by Oliver Cromwell, governed England. Boyle was a Royalist supporter but he kept his head down in a political sense and was thus able to pursue his scientific activities without hindrance.

The Beginning of Chemistry

In the 17th century the conventional divisions of science, which now designate a scientist as a chemist, physicist, biologist and so on, did not exist and scientists often spread their interests over the whole scientific spectrum. Isaac Newton, mentioned in the last chapter in relation to his activities as an alchemist — which might be thought of as pseudo-chemistry — is far better known for his work in gravitational theory, mechanics, optics and mathematics. So it was with Robert Boyle. Boyle's first noteworthy scientific accomplishment was the design and construction of an air pump. In 1650 a German scientist, Otto von Guericke, had invented an air pump that was capable of producing a good vacuum within a large enclosure. Von Guericke was the Mayor of Magdeburg and he designed an experiment, carried out in that city, to demonstrate the effectiveness of the vacuum he could produce. He constructed two copper hemispheres about half a metre in diameter (*Magdeburg hemispheres*) that could be joined together by an airtight grease seal to form a complete sphere. The sphere was evacuated using his pump and then a team of eight horses was harnessed to each hemisphere; for all their strength the horses were unable to separate the hemispheres (Figure 4.3).

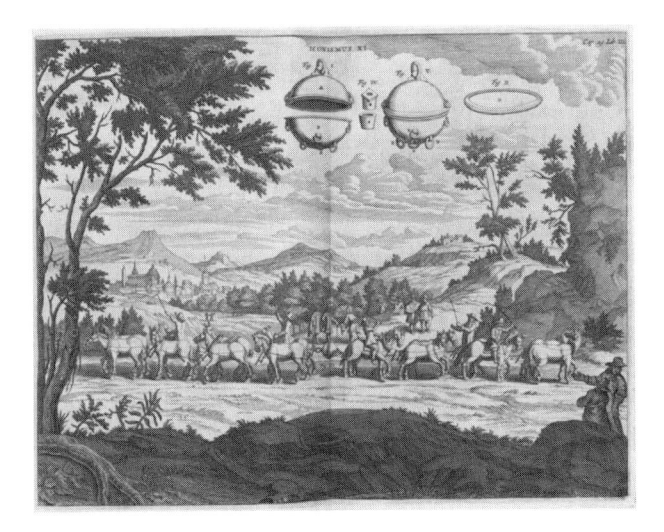

Figure 4.3 An engraving showing the Magdeburg experiment.

Boyle's air pump was completed in 1659, and with it he embarked on a series of experiments. These were reported in a treatise with the title *New Experiments Physico-Mechanicall, Touching the Spring of the Air and its Effects.* As a consequence of these experiments he formulated the law, generally known as *Boyle's law*, which states that '*For a fixed amount of gas at constant temperature the pressure of a gas is inversely proportional to its volume.*' This meant that if, for example, the gas is expanded to twice its previous volume then its pressure will halve and, conversely, if the gas volume is halved then the pressure will double. His interest in physical topics was considerable and varied and he also carried out experiments on the transmission of sound in air, on the measurement of the densities and refractive indices of materials, on crystals, on electricity and many other phenomena. However, for all his work in the area of what we would now call physics his first love, in a scientific sense, was chemistry.

Paracelsus was an early member of a group of 'alchemists', known as *spagyrists*, who had abandoned the search for the philosopher's stone, believed in the mercury-sulphur-salt set of basic principles, or elements, and concentrated on medical applications of chemistry. Boyle rejected their model of alchemy and retained a belief in the alchemical idea that one metal could be transmuted to another. While, in the 21st century, this may seem to be an irrational belief, in the state of knowledge of chemistry that existed in the 17th century there was some basis for it. For example, lead sulphide, in the form of the mineral *galena*, has a metallic appearance. When galena is heated the sulphur combines with oxygen in the air to give the gas sulphur dioxide leaving behind metallic lead that has properties very dissimilar to the apparently metallic galena. One metal has seemed to change into another by the process of heating. Again, galena often contains small amounts of silver sulphide and the heating then leaves behind a mixture of lead and silver that is indistinguishable in appearance and general properties from pure lead. However, on further heating the lead is converted into lead oxide that can be separated from residual metallic silver. This time the process has apparently turned lead into silver. There are other examples that, in the state of knowledge of the time, suggested that metals could be transformed from one to another and the hope that a base metal could be transformed into either silver or gold was not entirely irrational. Nevertheless, the activity of alchemists was regarded with great suspicion and, as already mentioned, the law introduced by Henry IV at the beginning of the 15th century that forbad 'multiplying gold and silver', essentially made the activities of alchemists illegal. Robert Boyle, and others, succeeded in getting this law repealed in 1689.

For all his belief in alchemy, Boyle nevertheless adopted a scientific approach in his chemical experiments, an approach he advocated in his first book on chemistry *The Sceptical Chymist*. His investigations were mostly concerned with the composition of chemical substances. He realised that there was a distinction between a compound, where two or more materials had come together to form a substance in which the ingredients had lost their identity, and a mixture, where the ingredients were still present in their original form and with their original properties. He was also formulating the idea of elements, basic components of chemical compounds that were themselves not produced by the combination of other substances. This idea is clearly conceptually similar to the early ideas of the four basic material elements — air, earth, fire and water — but with quite different elements. He was also the first to devise methods of detecting and recognizing the ingredients within mixtures and compounds, a process he called 'analysis', a term still used by chemists today. With remarkable prescience Boyle suggested that the chemical elements might themselves consist of other smaller bodies of various sizes and types but concluded that there was no way of determining what they were.

Modern chemistry owes much to the systematic approach of Robert Boyle. In 1663 the Invisible College formally became the *Royal Society of London for the Improvement of Natural Knowledge* under a charter granted by King Charles II and Boyle was a member of the first Council of the Society. The motto of the Royal Society '*Nullius in verba*' literally translates as '*Nothing in words*' and implies that statements alone are not enough but must be verified by factually based experiments. Robert Boyle exemplified this new approach to scientific endeavour.

Chapter 5

Modern Chemistry is Born

5.1 The Phlogiston Theory

Through his systematic scientific approach, Robert Boyle had begun the process of transforming alchemy into chemistry but, nevertheless, he was still an alchemist who believed that it was possible to change base metals into gold and silver. At the heart of alchemy was the concept of 'elements', a few entities with particular properties, combinations of which, in different proportions, could give all known substances. Converting one substance to another involved changing the mix of elements in some way.

At the beginning of the 18th century there was a prevailing belief in what was known as the *Phlogiston Theory*. This theory explained many chemical reactions in terms of a substance, or perhaps we should say 'element', of very low density called *phlogiston*, an element contained in combustible bodies and released by fire. Thus metals were thought to be phlogiston rich. When many metals are heated in the presence of air they form a powder, or brittle solid, known as a calx. This was thought to be due to phlogiston having being driven out of the metal but now we know it is due to the combination of the metal with atmospheric oxygen to form an oxide. Another substance thought to be rich in phlogiston was charcoal and, according to phlogiston theory, a calx heated with charcoal would absorb phlogiston from it and so be restored back to the metal. Indeed, when metallic oxides are heated with charcoal the metal is produced. Other patterns were established. For example, when a combustible material burns then it loses weight because the phlogiston within it is expelled. In particular, charcoal was thought to be almost pure phlogiston so when it was burnt very little residue was left. As a final example of a phlogiston-based explanation, when combustion of a material takes place in an enclosure then the burning will cease once the air becomes saturated with phlogiston and can absorb no more. There was a certain self-consistency in the phlogiston model but it also had problems. A loss of phlogiston by charcoal involved a great loss of mass whereas when phlogiston was lost from a heated metal the resultant calx was *heavier* than the metal. However, belief in the phlogiston theory was so strong that its supporters believed that any explanation that was necessary to preserve its integrity had, perforce, to be true. Thus, since driving phlogiston out of a metal gives a calx that is heavier than the original metal then this was taken to show that the phlogiston in a metal, by contrast with that in charcoal, had a *negative mass*. Such an explanation may seem rather absurd now but, at the time, it seemed plausible and temporarily preserved the viability of the phlogiston theory (Figure 5.1).

5.2 Joseph Priestley

A very unlikely contributor to the advancement of chemistry during the 18th century was an English clergyman, Joseph Priestley (Figure 5.2). He was a man of many talents and his studies and work covered the fields of philosophy, theology, political science, physics and chemistry. He also found the time to learn a whole raft of languages in addition to the Latin, Greek and Hebrew he studied at school — French, German, Italian, Chaldean (an Aramaic-rooted

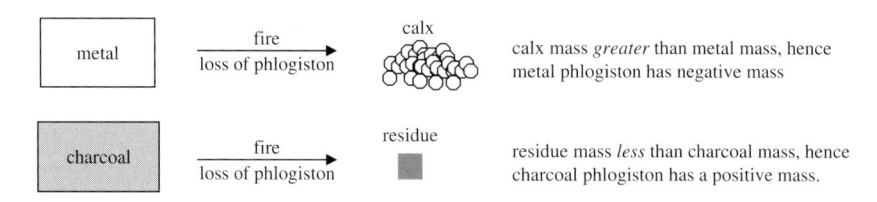

Figure 5.1 The mass associated with phlogiston from different experiments.

Figure 5.2 Joseph Priestly (1733–1804).

language spoken in some parts of Iraq), Syriac (an ancient language used by some Christian groups in the Middle East) and Arabic.

Many of his non-scientific activities brought him into conflict with the establishment. In religious terms he was a Rational Dissenter. Dissenters were a religious group that had separated itself from the Church of England and were discriminated against by an Act of Parliament that prevented them from holding political office, serving in the army or navy or attending the universities of Oxford or Cambridge — the only universities in England at that time. Rational Dissenters went further, proclaiming that there should be no links between state and church, especially financial, and also that belief should be based on personal choice, rational thought and science rather than on tradition and dogma. Against the general trend of opinion in England, Priestley was also a strong supporter of the French revolution, believing it to be the rightful deposition of an unjust monarch.

As was true for many scientists of his time, Priestley's investigations spanned the boundaries of what today would be regarded as separate sciences. He studied electricity and showed that charcoal (carbon) was a conductor thus disproving the current belief that only water and metals were conductors. He also showed that there were varying degrees of conductivity in a range of different materials giving a continuum from insulators through to highly conducting metals. In this work on electricity he received active help and encouragement from Benjamin Franklin, the eminent American scientist, whom he met on one of Franklin's periodic visits to London. It was as a result of his work on electricity that Priestley was elected a Fellow of the Royal Society in 1766. Another area that interested him was that of vision, optics and colour. However, his most significant, and best-known, contributions to science were his studies related to gasses that, in those days, were referred to as 'airs'.

It must be stressed that in the mid-18th century the idea of chemical elements and the way they combined to form compounds was not understood. Thus different gasses were described by names that indicated their nature and properties. Before Priestley began his work on gasses there were only three that were explicitly recognized — atmospheric air, carbon dioxide and hydrogen. When Priestley moved to Leeds in 1769 his home happened to be close to a brewery that was a copious source of carbon dioxide, a by-product from the fermentation process that produced the alcohol in beer. He discovered how to impregnate water with carbon dioxide and wrote a pamphlet describing the process in 1772. The fizzy drink so produced was mistakenly thought to be a possible antidote against scurvy, a terrible scourge of those making long sea journeys in those days. Although this turned out not to be a property of the drink, in 1783 a Swiss watchmaker, Jean Jacob Schweppe, transformed this process into a commercial venture by producing carbonated mineral water. In 1790 he set up his carbonated water (soda water) drinks industry in England where, in one guise or another, it has existed ever since.

In a paper published in 1775 in *The Philosophical Transactions of the Royal Society*, Priestley described an experiment in which, with the aid of a 12-inch burning glass, he heated up what we now know to be red mercuric oxide and produced a remarkable gas — one that we now know to have been oxygen. He described its properties thus:

> 'This air is of an exalted nature. A candle burned in this air with an amazing strength of flame; and a bit of red hot wood crackled and burned with a prodigious rapidity, exhibiting an appearance something like that of iron glowing with a white heat, and throwing off sparks in all directions. But, to complete the proof of the superior quality of this air, I introduced a mouse into it; and in a quantity, which had it been common air, it would have died in about a quarter of an hour. It lived at two different times, a whole hour, and was taken out quite vigorous.'

Priestley also isolated and described many other gases including — with his name for them in parentheses — nitric oxide (nitrous air), nitrogen dioxide (red nitrous vapour), nitrous oxide (diminished nitrous air, later called 'laughing gas') and ammonia (alkaline air), but without any idea of their true chemical nature.

Priestley interpreted his experiments with oxygen, involving candles, glowing embers and mice, in terms of the 'quality' of the air before and after the experiment. According to the phlogiston theory, air became incapable of supporting combustion after it became saturated with phlogiston. This phlogiston-saturated air was of the worst quality. Normal air contained some phlogiston but was not saturated so that some burning could take place in it. It was of intermediate quality. The best air was that coming from the heating of mercuric oxide, and was designated as 'dephlogisticated air' — air that was completely free of phlogiston and so would support extremely vigorous combustion. For all the importance of his work on oxygen, Priestley never understood its true relationship to combustion and respiration and he certainly had no concept of the formation of oxides. He was a strong supporter of the phlogiston theory for the whole of his life.

Priestley was often publicly criticized and put under pressure for his defence of the French revolution. In 1791, on the second anniversary of the storming of the Bastille when he was living in

Birmingham, a mob destroyed his house and all his possessions therein, including his library, after which he left Birmingham, never to return. From Birmingham he moved to Clapton, near Hackney, then a small rural community not far from London, where he taught at the local college. However, in 1793, after the execution of Louis XVI and the outbreak of war between England and France, the social pressure on him became too great to bear and he left for America where he lived for the remainder of his life.

5.3 Antoine Lavoisier

During the latter part of the 18th century a new scientific star was in the ascendant, through whose work the phlogiston theory would eventually receive its *coup de grâce*. Antoine Laurent Lavoisier (Figure 5.3) was the French scientist whose right to bear the title 'Father of Chemistry' is stronger than that of any other individual. His great contribution to chemistry was to make it quantitative by making accurate measurements — especially changes of mass of the components of his experiments, to which

Figure 5.3 Antoine Laurent Lavoisier (1743–1794).

end he constructed a balance capable of measuring to an accuracy of half a milligram. His wife, Marie-Anne, whom he married when she was just 13 years old, supported him in his work. She helped him both with his experiments and also by translating scientific papers to be written in English.

Lavoisier started his research, like all other scientists of his time, accepting the phlogiston theory of combustion and in 1773 he started a series of investigations to analyse the products of combustion in different circumstances. In this work he was certainly stimulated by a visit from Priestley in which Priestley is said to have imparted the knowledge of how to produce 'dephlogisticated air'. To understand the language of the time in relation to gasses one needs to interpret the then-current terminology in terms of what the gasses were. Thus:

'dephlogisticated, or pure air'	≡	oxygen
'dead air'	≡	nitrogen
'fixed air'	≡	carbon dioxide
'inflammable air'	≡	hydrogen
'common air'	≡	normal atmospheric gas.

In one experiment Lavoisier suffocated a bird in a glass enclosure (a lack of squeamishness was a characteristic of the time) and then showed that a lighted candle introduced into the enclosure was extinguished. This established a strong link between combustion and respiration and suggested that the residual air left over from suffocating a bird was similar to that when a candle was extinguished. Next Lavoisier tested the air left by an extinguished candle. He found that this air, shaken up with water, lost one-third of its volume — we now know that the 'fixed air' (carbon dioxide) is produced by combustion or respiration combined with the water to form carbonic acid, leaving behind 'dead air' (nitrogen). In another series of experiments he burned a candle in different mixtures of 'common air' and 'pure air' in an enclosure situated over mercury so that none of the 'fixed air' would be lost by dissolving in water. When caustic alkali was introduced into the enclosure the 'fixed air' reacted with it and was removed and the consequent reduction in the total volume of gas was measured.

In this way Lavoisier showed that more 'fixed air' was produced when there was more 'pure air' in the enclosure, which suggested that the result of combustion was the conversion of 'pure air' to 'fixed air', that is in modern terminology that oxygen was being converted into carbon dioxide. Not all the gas in 'common air' was converted to 'fixed air' so it was clear that 'common air' was a mixture of airs; the air left in the enclosure after combustion and the removal of 'fixed air' was 'dead air', now known as nitrogen. This description of Lavoisier's gas experiments sounds complicated because it has been expressed in the language of the time. If we made an equivalent statement in modern terms that a mixture of oxygen and atmospheric air (approximately 20% oxygen and 80% nitrogen) would produce more carbon dioxide than the same quantity of atmospheric air and leave a lesser residue of nitrogen then that becomes far more understandable.

In the experiments with gasses Lavoisier used the measurement of volumes but in other combustion experiments he used his scales accurately to measure masses. In one early experiment he burnt a material called *pyrophor*, produced by the distillation of faeces — the component *pyro* (from the Greek word *pur* meaning 'fire') in the name related to its tendency to spontaneously burn when exposed to air. This material was a mixture, predominantly of phosphorus, sulphur and carbon and Lavoisier noted that the product of the combustion was greater than the original mass. Lavoisier must have been uncomfortable with the idea of the negative mass of phlogiston in metals but now he had found another completely non-metallic material that gained mass on combustion. He wondered — was the combustion process not actually removing phlogiston of negative mass but rather adding something of positive mass?

Another crucial series of experiments involved heating a mixture of charcoal and calxes in a sealed container. Lavoisier found that the mass of the container was completely unchanged by the heating process but when the container was opened 'fixed air' escaped from it and metal remained in the container. From this and other experiments Lavoisier formulated the idea of mass conservation — the principle that when a chemical reaction takes place the total mass of the original reactants equals that of the final products.

Although Priestley made the initial discovery of 'dephlogisticated air' there can be no doubt that Lavoisier actually recognized it for what it was. He confirmed that when a metal was heated in 'common air' some part of it combined with the metal to give a calx and that was the component that was the same as 'pure air'. What was left of the 'common air' was 'dead air.' From this it could be deduced that 'common air' was just a mixture of 'pure air' and 'dead air' or, as we would say now, atmospheric air is a mixture of oxygen and nitrogen. There is some dispute about who actually discovered oxygen; Lavoisier did tend to use the results of other people's work without acknowledging their contribution. In this case it is probably fair to divide the credit equally between Priestley and Lavoisier. Priestley's claim may have priority but without Lavoisier's work the discovery of oxygen would have been of little value.

In 1789, the year of the French revolution, Lavoisier published *Traité Élémentaire de Chimie*, the first proper chemistry textbook. In it he presented the principle of the conservation of mass and argued against the phlogiston theory. He also introduced the idea of chemical elements and gave names to them, giving as examples hydrogen, oxygen, nitrogen, phosphorus, mercury, zinc and sulphur. However, he did get some things wrong — light and caloric (heat) were included in his list of elements. With Lavoisier, chemistry emerged from its alchemical roots and became a well-ordered scientific discipline.

Priestley's life was greatly affected by his espousal of the French revolutionary cause but Lavoisier's life was affected by the revolution even more. Prior to the revolution he had been a tax collector and quite an important man in the French establishment of the time. In 1794, during the Reign of Terror he was denounced as a traitor, tried, found guilty and guillotined. Less than two years after his death the French government realised that in executing this brilliant scientist they had committed a grave error and an injustice and so they posthumously exonerated him. However, like the suffocated bird, he could not be restored to life and the work of a great scientist was tragically cut short at the early age of 50.

Chapter 6

Nineteenth Century Chemistry

6.1 Chemistry Becomes Quantitative

By the end of the 18th century Lavoisier had established the idea that chemistry was based on a large number of elements and not just a few — such as mercury, sulphur and salt that conferred certain properties on substances — as proposed by Paracelsus. During the 19th century, starting from this concept and by a series of gradual steps taken by eminent scientists of many nationalities, chemistry as the modern subject we know today steadily evolved.

The earliest work in this development of modern chemistry began in the late 18th century. Other scientists at that time were following the path blazed by Lavoisier of carrying out quantitative work. There was an understanding of the idea of bases — alkaline materials, which reacted with and neutralized acids. Two German chemists, Karl Wenzel (1740–1793) and Jeremias Richter (1762–1807), carried out a series of experiments in which they determined the masses of different bases that neutralized a fixed mass of acid. What they found is that if the ratio of the masses of two bases required to neutralize acid A was R then the same ratio, R, of the base masses was found when neutralizing a different acid, B. This quantitative approach to chemistry they called *stöichiometrie* (in English, stoichiometry) and it established the principle that there were underlying quantitative laws that governed the masses of materials taking part in chemical reactions. What Wenzel and Richter had established for base-acid reactions was taken further by the French scientist Joseph Proust (1754–1826) and applied to other kinds of reaction. He formulated the rule that when materials react the ratio of the masses of the reactants is fixed and cannot be arbitrarily or continuously varied. This law, called either *Proust's law* or *the law of definite proportions*, was a new and important concept, although to those brought up in the modern scientific age this law seems to be axiomatic and obvious. However, to the scientists at the end of the 18th century, not long emerged from the mysticism of the alchemical era, it represented a watershed in thinking.

6.2 John Dalton

An extremely important contributor to this steady advance of chemistry was the Englishman, John Dalton (Figure 6.1). In the tradition of those that preceded him he was a scientist of wide range, spanning many of the distinct scientific disciplines that we recognize today. Dalton was born into a Quaker family, the son of a weaver, in the Lake District of northern England. Like Joseph Priestley he was a Dissenter and so was not allowed to attend a university. Consequently Dalton's education was acquired in piecemeal fashion from various individuals he met in his early life. One of these, Elihu Robinson, a

Figure 6.1   John Dalton (1766–1844).

fellow Quaker, was a meteorologist and this subject was an early interest for Dalton and provided the material for his first scientific paper, *Meteorological Observations and Essays*, published in 1793. His next major interest was colour vision, prompted by the fact that he suffered from a particular form of colour blindness in which there is an inability to distinguish colours in the green-to-red part of the spectrum. In 1794 he moved to Manchester where he was elected to the Manchester Literary and Philosophical Society, a highly respected learned society. Shortly afterwards he contributed an article to the Society's journal on the topic of deficiencies in colour vision, which he accurately described but for which he gave an incorrect explanation. So influential was this article that the most common form of red-green colour blindness is now known as Daltonism.

In 1801 Dalton gave a series of talks, later published in 1802, on the pressures exerted by mixtures of gasses, the pressure of various gasses and vapours, including steam, at different temperatures and on the thermal expansion of gasses at constant pressure. From these experiments he expressed the view that given a low enough temperature and under a sufficiently high pressure any gas could be liquefied, something we now know to be true. He also expressed, although not in language that seems very clear today, the gas law presented by Joseph Gay-Lussac (1778–1850) in 1802 giving the relationship between the volume of a gas and its temperature at constant pressure. Gay-Lussac acknowledged that his compatriot, Jacques Charles (1746–1823) had previously discovered, but not published, this relationship so it is now generally known as Charles's law.

John Dalton's notebooks, found in the archives of the Manchester Literary and Philosophical Society, indicate that during his studies of gasses he was beginning to turn his mind towards the concept of atoms. The idea of an atom had first arisen with Leucippus and Democritus (Chapter 2) but their concept of an atom was an indivisible unit of any substance — for example, an atom of stone or an atom of wood. Lavoisier had introduced the idea of chemical elements, as we know them today, basic substances that combined together to form all other materials but which themselves could not be formed from anything else. Dalton's atoms brought together these two concepts.

His atoms were the indivisible units of the individual elements; these atoms could combine together to form chemical combinations, or *compounds*. The idea of elemental atoms led him to *the law of multiple proportions*. For example, Dalton noted that, 'The elements of oxygen may combine with a certain portion of nitrous gas (nitrogen) or with twice that portion, but with no intermediate quantity.' There is a general belief that Dalton was led to his idea of atoms by consideration of the law of definite proportions, previously given by Proust. However, it is clear from Dalton's notes that the reverse was true; his atomic theory was of a purely conceptual nature and from it the law of multiple proportions could be derived as a natural consequence. Dalton described compounds involving two elements in terms of the number of atoms they contained. Thus a binary compound consisted of atom A plus atom B, a tertiary compound as atom A plus two atoms B.

Dalton expressed the view that the atoms of one element are all identical and different from those of any other element. He also stated that they could be distinguished by their relative weights, and on the basis of assumptions about the numbers of atoms forming particular compounds, and the actual weights of the materials involved in reactions, he calculated some relative atomic weights based on a scale in which the weight of a hydrogen atom is unity. Since he made errors in some of his assumptions about chemical formulae — he assumed that water consisted of one atom of hydrogen with one atom of oxygen (actually two hydrogen to one oxygen) and ammonia was one atom of hydrogen with one atom of nitrogen (actually three hydrogen to one nitrogen) — his relative weights were incorrect, but the basic idea of atomic weights was established. Dalton's contribution to the concept of atomic weights is recognized by the use of the unit *dalton* (Da) as one atomic mass unit, taken nowadays as one-twelfth of the mass of the most common carbon isotope,[c] very closely the mass of a hydrogen atom. The idea that chemistry involved the linking of atoms of individual elements to form compounds established the basis of modern chemistry.

---

[c] The atoms of many elements have different forms called *isotopes*. The meaning of this term is explained in Chapter 15.

6.3 Amedeo Avogadro

Dalton published his atomic theory in 1808 in the book, *New System of Chemical Philosophy*. Shortly afterwards Gay-Lussac showed that the ratios of the volumes of gasses that gave a chemical reaction and the volume of the product, if gaseous, can be expressed in small whole numbers if all the gasses were at some standard pressure and temperature. The work of Dalton and Gay-Lussac led directly to the next advance in the understanding of chemistry made by the Italian, Amedeo Avogadro (Figure 6.2) who was actually a physicist and mathematician. Avogadro clarified the idea that a gas consisted of molecules that themselves consisted of combinations of atoms. He put forward the law that bears his name that states that: the relative masses of the same volume of different gasses at the same pressure and temperature are the same as their relative molecular weights. From this it follows that the relative molecular weight of a gas can be found by measuring the mass of a known volume of a gas at a standard, or known, temperature and pressure.

One of the consequences that came from measurements of the masses of gasses and by the application of Avogadro's law is that some gasses contain molecules that are a combination of two atoms of the

Figure 6.2 A drawing of Amadeo Avogadro (1776–1856).

same kind — for example, two hydrogen atoms or two oxygen atoms. This was something that the current ideas of chemistry would not allow. Dalton's model of an atom was that it was surrounded by 'an atmosphere of heat' and that the atmospheres of like atoms would repel each other. This conclusion was supported by the work of another scientist, the Swedish chemist Jöns Jacob Berzelius (1779–1848), who has the distinction of having invented much of the chemical terminology in use today — for example, representing atoms by letter symbols, such as O for oxygen and H for hydrogen. Berzelius had carried out experiments on electrolytic decomposition in which, by the passage of an electric current through liquids, molecules were broken up, with some elements designated as *electropositive* going to the negative terminal, the *cathode*, while other elements designated as *electronegative* went to the positive terminal, the *anode*. Thus the electrolysis of water gives electropositive hydrogen released at the cathode and electronegative oxygen released at the anode. From this work Berzelius concluded that elements were either electropositive or electronegative so that like atoms would automatically repel each other and hence that a molecule consisting of two like atoms could not form. He postulated that molecules formed because of the attraction of atoms of different polarity — one electropositive and the other electronegative. This idea was superficially attractive, and generally accepted, so Avogadro's law, which requires some elemental gasses to be in the form of pairs of similar atoms forming a molecule, was never fully accepted during his lifetime.

A much better understanding of the way that atoms combine to form molecules came about shortly after this initial rejection of Avogadro's law. During the early part of the 19th century there was considerable interest in the chemistry of organic compounds — compounds containing carbon. A notable French organic chemist, Jean Baptiste André Dumas (1800–1884) together with his student, Auguste Laurent (1807–1853) challenged the electrochemical model of molecular formation advanced by Berzelius. In one notable experiment Dumas prepared trichloroacetic acid by substituting three hydrogen atoms in acetic acid by three atoms of chlorine. Laurent noted that substituting the electropositive hydrogen atoms with electronegative chlorine gave a compound with very similar properties, from which he concluded that the properties depended mainly on what had remained unchanged in the molecule — everything other than the hydrogen or chlorine atoms. He also noted that the hydrogen and chlorine atoms had been equally able to attach themselves to this unchanged component regardless of their different polarity. He concluded that molecules consisting of pairs of like atoms *can* form and hence that all chemical bonding does not occur in the way that Berzelius suggested. Because of this argument Avogadro's law was eventually accepted.

Avogadro's contribution to chemistry is recognized through the physical constant *Avogadro's number*, $6.022\,14 \times 10^{23}$. The *molecular weight* of a chemical compound is the mass of one unit (molecule) of the compound in dalton units. One *mole* of a substance is the amount of that substance that will give the molecular weight in grams and Avogadro's number is the number of molecules in one mole. Thus the atomic weight of sodium is 23, which is to say that a sodium atom has 23 times the mass of a hydrogen atom, and $6.022\,14 \times 10^{23}$ atoms of sodium have a total mass of 23 grams.

The determination of this number, by the Austrian chemist, Josef Loschmidt (1821–1895) relates to some other very fundamental work being carried out in the 19th century. This was in the field of *statistical mechanics*, that is the application of probability theory to the behaviour of large systems of particles — for example, the molecules in a gas. The leading figures in this field were the German physicist, Ludwig Boltzmann (1844–1906) and the English physicist Lord Kelvin (1824–1907). To determine Avogadro's number, sometimes called the *Loschmidt number*, Loschmidt used a combination of theory from statistical mechanics and experimental work in which he measured the volume of liquefied gasses. The value he obtained for Avogadro's number was not very precise but it was of the right order of magnitude.

6.4 The Concept of Valency

Auguste Laurent, the one-time student of Jean Baptiste Dumas, in association with another French chemist, Charles Frédéric Gerhardt (1816–1856), established a new notation for chemical formulae that systematized what had previously been fairly chaotic. One important

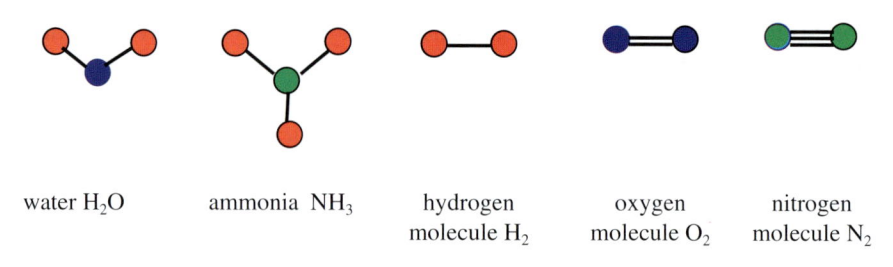

Figure 6.3 Some simple molecules illustrating valence. ● hydrogen ● oxygen ● nitrogen.

concept that came out of this work was that of *valency*, a term that describes the power of an atom to combine with other atoms. Thus hydrogen has a valency of one, oxygen two and nitrogen three so that, for example, an oxygen atom, with a valency of two, can combine with two hydrogen atoms, each with a valency of one, to give a molecule of water. Similarly a nitrogen atom, with a valency of three, can combine with three atoms of hydrogen to give a molecule of ammonia. It is not certain that these early chemists actually believed in the real existence of atoms and molecules but, together with the concept of valency, they did regard them as giving conceptually useful models for describing compounds and even predicting what compounds might exist. As simple examples Figure 6.3 shows representations of water and ammonia and also hydrogen, oxygen and nitrogen molecules. Between the oxygen atoms there is a *double bond* because oxygen has a valency of two and a *triple bond* links the nitrogen atoms in a nitrogen molecule. All other bonds shown are single and the number of bonds connected to each atom indicates its valency.

With the emergence of these new concepts chemistry, which in its beginnings had something of the nature of a culinary art, was taking on the characteristics of a formal science.

6.5 Chemical Industry is Born

Although chemistry as a scientific subject did not exist in ancient times the use of chemicals for a whole range of applications was widespread.

The use of natron by Egyptian priests to dry out bodies for mummification has already been mentioned. Another common chemical in frequent use in the ancient world was urine, which was used for curing leather. In ancient Rome large containers were distributed around the streets of the city into which people could urinate — public toilets of a sort but perhaps too public for modern taste. The contents were collected by tanners and allowed to stand for a time during which the urine transformed into ammonia that was the actual tanning agent. In the city of Fez in Morocco, which is famed for the quality of its leather, this practice of using urine for tanning persists to the present day.

It is difficult to define a time when it could be said that a chemical industry first came about. However, as an early example, we can take the origin of the alum industry in England. There is a whole class of chemical compounds, all different, called 'alums' but the name *alum*, without further qualification, is applied to the compound potassium aluminium sulphate — as a chemical formula $KAl(SO_4)_2 12H_2O$, where K is potassium, Al is aluminium and S is sulphur. The other symbols we have met previously. Alum has many uses but the major use was as a *mordant* in the dyeing of textiles, particularly wool. If woollen yarn is immersed in a natural dye, derived from some vegetable or animal source, then it takes up the colour of the dye but then the colour washes out when the yarn is wetted. The Romans discovered that if the yarn was first immersed in an alum solution before being soaked in the dye then the colour would be permanently retained. Sources of alum were comparatively rare so the marketing of the substance was extremely profitable. In the 15th century the production and sale of alum became a monopoly of the Catholic Church, centred on the Vatican, and this presented a severe problem to the English woollen industry in 1533 when Henry VIII adopted Protestantism after his dispute with the Pope about divorcing Catherine of Aragon. However, it turned out that there are vast deposits of shale at Ravenscar on the coast of North Yorkshire from which alum could be produced. The process involved heating the shale in huge bonfires for several months after which the shale turned a reddish colour. Then the reddened rock had to be treated with urine (useful stuff!) to extract the alum; since the region was sparsely populated this required importing large quantities of urine which came in by ship from as far away as

London. It is claimed that the first public toilet in Hull, the major Yorkshire port, was set up to provide urine for alum extraction. The alum industry of Ravenscar flourished from the early 17th century through to the middle of the 19th century.

In the 19th century chemistry took off as an activity of great commercial importance. Until the mid-19th century all textiles had been coloured with natural dyes, usually but not always of vegetable origin. In 1856 a young English chemist, William Perkin (1838–1907) was attempting to prepare a synthetic form of quinine, a material derived from the bark of the cinchona tree which grows in South America and which is a treatment for malaria. Perkin began his synthesis with the compound aniline and ended up with a rather sticky black substance. He found that, by the use of alcohol as a solvent, he could extract an intense purple colour from the black goo. This was commercialized as the synthetic dye *mauveine* and so began an important industry for producing synthetic dyes, many of them aniline derivatives. Although this serendipitous breakthrough in producing dyestuffs was made in England, it was Germany that eventually became the powerhouse in this industry and by the end of the 19th century German companies were dominant in the field. Later a chemical conglomerate *Interessen-Gemeinschaft Farbenindustrie AG* (*Syndicate of dyestuff corporations*), generally known as IG Farben, was formed that became a massive general chemical company.

The latter half of the 19th century and the first half of the 20th century was a period of the worldwide dominance of chemistry in an industrial sense. Apart from the great German chemical industry there were large companies and conglomerates of companies in most industrialized countries, producing fertilizers, explosives, dyestuffs, paints, non-ferrous metals, pharmaceuticals and many other products. Switzerland, a small country, actually started making pharmaceuticals in the mid-18th century and is still a major centre of pharmaceutical production.

6.6 Bringing Order to the Elements

During the 19th century the tally of known elements steadily increased, as did knowledge of their chemical properties. A German chemist, Julius Lothar Meyer (1830–1895), in 1864 and an English chemist, John Newlands (1837–1898), in 1865, noticed that there seemed to be patterns linking the atomic weights of groups of elements having similar chemical properties but their descriptions of these patterns were not very convincing. For example, Newlands did not make any allowance in his table for elements still undiscovered and sometimes had to put two elements in the same location of the table to preserve the pattern he was advocating. To give an example of what Newlands reported in 1865 the following are the first 17 elements known to him, written down in columns in order of atomic weight, with the atomic weight in brackets. What Newland noticed is that elements in the same row have similar properties. However, that pattern broke down as more elements were added.

hydrogen (1)
lithium (7)   sodium (23)   potassium (39)
beryllium (9)   magnesium (24)   calcium (40)
boron (11)   aluminium (27)
carbon (12)   silicon (28)
nitrogen (14)   phosphorus (31)
oxygen (16)   sulphur (32)
fluorine (19)   chlorine (35)

Then between 1868 and 1870 a Russian chemist, Dmitri Mendeleev (Figure 6.4), published the two volumes of his textbook *Principles of Chemistry* in which he classified elements according to their properties and described the pattern he perceived in what he called the *Periodic Table*. The initial table he produced was the following:

Cl	35.5	K	39	Ca	40
Br	80	Rb	85	Sr	88
I	127	Cs	133	Ba	137

Figure 6.4   Dmitri Mendeleev (1834–1907).

The elements in each column have similar chemical properties and the differences in atomic weights are similar. The first column contains chlorine, bromine and iodine and the differences of atomic weight are 44.5 and 47. The second row contains potassium, rubidium and caesium and the differences in atomic weight are 46 and 48. The final row contains calcium, strontium and barium with differences of atomic weight 48 and 49. Starting from this small table Mendeleev added new rows and columns until he had built up a complete version of the Periodic Table.

At the time that this first version of the table was prepared there were many unknown elements and so, to preserve the rationale of the table, Mendeleev had to leave several gaps in it. He predicted that elements would be found to fill these gaps and was able to say what their chemical properties would be and also to give estimates of their atomic weights. He correctly predicted the discovery of the elements germanium, gallium and scandium in this way — although his suggested names for them were different.

We shall see later that the key to providing a logical Periodic Table is not to be found in atomic weights but in another property that atoms possess, one that is correlated with their atomic weights. However, in the light of the knowledge of the time, Mendeleev's Periodic Table represented a considerable advance in the process of making chemistry into a well-founded science.

Chapter 7

Atoms Have Structure

7.1 Michael Faraday

The Greek philosophers, who sought simplicity and harmony in describing the world, were well pleased with their four elements, combinations of which produced all possible substances. However, chemists in the 19th century were finding ever more chemical elements and Mendeleev's Periodic Table, with its many gaps, promised even more discoveries in the future. The potential number of elements was still not very big but, even so, our Greeks would have been unhappy that the number had grown so large. In this chapter we shall describe work that began in the first half of the 19th century and gave the first discovery that, with others, would once again have restored contentment to the Greeks of old.

The invention of the vacuum pump by Otto von Guericke, and its subsequent improvement by Robert Boyle (Chapter 4), stimulated scientists in the years that followed to carry out numbers of vacuum-based experiments — although in practice the 'vacuum' was actually a space with some air in it at very low pressure. In 1838 the English scientist, Michael Faraday (Figure 7.1), passed an electric current through an evacuated tube and observed that within the tube there was a very pale band of light stretching from very close to the cathode (negative terminal) right up to the anode (positive terminal). The region close to the cathode where there was no light is now

Figure 7.1   Michael Faraday (1791–1867).

called the *Faraday dark space*, or sometimes the *Crookes dark space* after a later worker on this phenomenon. It was deduced that something was passing along the tube to produce this band of light but nobody had any idea what that something could be.

Michael Faraday was one of the most extraordinary scientists of his, or indeed any, time. He had little formal education, knew very little mathematics, but had great strength as an experimentalist. He attended lectures at the Royal Institution in London and made himself known to its Director, Humphrey Davy (1778–1829), who gave the lectures. Davy was impressed by the young man and eventually appointed him as his assistant — a rather menial role in those days. Faraday's subsequent pioneering work in many fields of chemistry and physics quickly earned him recognition in the scientific community. He was elected a Fellow of the Royal Society at the age of 32 and one year later, in 1825, he succeeded Humphrey Davy as Director of the Royal Institution. In addition to his vacuum tube experiments, Faraday carried out another kind of experiment that eventually had consequences in elucidating the nature of matter. These experiments involved the process of electrolysis, used by Berzelius to show that some elements were electropositive and others electronegative (Chapter 6). In 1857, as a result of his investigations of electrolysis, Faraday concluded that 'The amount of bodies which are equivalent to each other in their ordinary chemical action have equal quantities of electricity naturally associated with them'. The quantity of electricity passing through an electrolytic cell is the product of current and time; a current of 1 *ampere* (symbol A) flowing for 1 second corresponds to the flow of a quantity of electricity, or *charge*, equal to 1 *coulomb* (symbol C). As an example of what this conclusion means, if a current is passed through a solution of copper chloride then copper is deposited at the anode and, similarly, for a solution of silver chloride, silver is deposited at the anode. Copper and silver both have a valency of 2 and, for an equal quantity of electricity (number of coulombs) the masses of copper and silver deposited will be in the ratio 63.5:65.4, the ratio of their atomic weights.

A summary of Faraday's discovery from electrolysis can be expressed as follows. For any monovalent element (of valency 1) the quantity of electricity required to deposit one mole (Avogadro's number of atoms) is 96,485 coulombs. This quantity of electricity is now called the *faraday* — symbol F. To deposit one mole of a divalent element (of valency 2) requires a quantity of electricity 2F and, in general, to deposit one mole of an element of valency v requires a quantity of electricity vF. Contained within this conclusion is the idea that an atom has associated with it a quantity of electricity proportional to its valency. The significance of this conclusion only became apparent after further work in the years to come.

7.2 The Nature of Cathode Rays

The nature of what was moving along the vacuum tube when a current was passed through it was a matter of great interest to the scientific community in the latter half of the 19th century. Was it radiation of some kind or was it a stream of particles? If it was radiation then what kind of radiation and if particles what kind of particles? During the period between the mid-1850s and the end of the 1860s there were a number of significant investigations of this phenomenon carried out by German physicists. In 1857, using a much improved vacuum pump able to achieve much lower pressures, Heinrich Geissler (1814–1879) found that the residual air in the tube glowed quite brightly. Substituting other gasses for air in the tube gave bright lights of different colours and the *Geissler tube* is the basis for the decorative displays now used for shop signs and advertising — for example, neon signs. A year later Julius Plucker (1801–1868), discovered that the band of light could be deflected with a magnet and in 1865, using an improved Geissler tube at an even lower pressure, he found a glow over the walls of the tube that was also affected by an external magnetic field. Finally in 1869 Johann Wilhelm Hittorf (1824–1914) found that sometimes the glass at the anode end of the tube gave a fluorescent glow. This fluorescence was taken as being due to radiation — *cathode rays* as they came to be known — starting from the cathode and travelling towards the anode where they caused the fluorescence.

The general conclusion from this work, and the one favoured by the German scientific community at that time, was that cathode rays were some form of radiation. However, in 1871 an English telegraph engineer, Cromwell Fleetwood Varley (1828–1883), published a paper in which he suggested that cathode rays consisted of charged particles and cited the behaviour of the stream of light in a magnetic field to support this view. A charged particle moving at right angles to a magnetic field is deflected in a direction perpendicular to both the direction of motion and that of the field. The direction of the deflection indicates whether the charge is positive or negative and, on the assumption that the particles were moving from the cathode to the anode, the deflection indicated that the charge was negative. An English chemist, William Crookes (1832–1919), suggested that cathode rays might be molecules of the gas in the tube that had picked up a negative charge from the cathode and were then propelled towards the positive anode. He also designed a new type of tube specifically to investigate the fluorescent glow at the anode end of the tube. The general form of the *Crookes tube* is illustrated in Figure 7.2. The anode, in the shape of a cross, is opaque to the cathode rays and its shadow

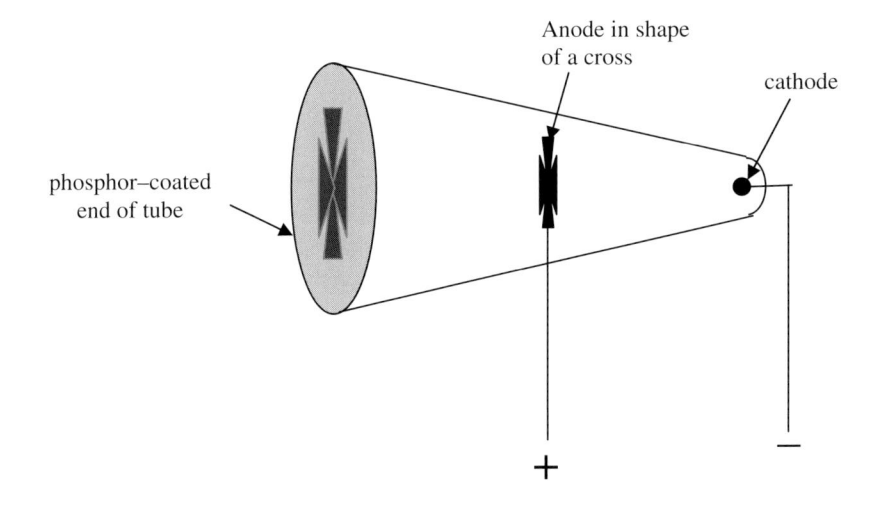

Figure 7.2 A schematic Crookes tube.

image is seen on the phosphor-coated anode. A magnet placed in the vicinity of the tube deflects the image on the screen.

Another line of investigation in the latter half of the 19th century was into the nature of electric charge. Faraday's work on electrolysis had shown that the amount of charge required to liberate one mole of an element of unit valency was one faraday, 96,485 coulombs. In 1874 George Johnstone Stoney (1812–1911), an Irish physicist, divided the faraday by Avogadro's number to give a value of about 10^{-20} C per univalent atom. This gave a basic electric charge associated with, say, a hydrogen atom, and the electric charge associated with an atom of valency v would then be v times as great since it would require v times as much charge to produce the same number of atoms in an electrolysis experiment. Stoney gave this unit of electric charge the name 'electrine' but in 1891 he changed it to 'electron'. A leading German physicist, Herman Ludwig von Helmholtz (1821–1894), in a lecture given at the Royal Institution in London in 1881, from a slightly different point of view, also suggested that there was a basic unit of charge associated with atoms. The form in which he gave this conclusion was "Now, the most startling result of Faraday's law is perhaps this. If we accept the hypothesis that the elementary substances are composed of atoms, we cannot avoid concluding that electricity also, positive as well as negative, is divided into definite elementary portions, which behave like atoms of electricity."

To test the hypothesis that cathode rays may be charged particles, as suggested by Varley, Heinrich Hertz (1857–1894), who began his career in physics as a student of Helmholtz, set up an experiment in 1883 in which a beam of cathode rays passed between two charged metal plates between which an electric field was established. When passing through an electric field, positively charged particles are deflected in the direction of the field and negatively charged particles in the opposite direction, but Hertz detected no deflection at all. From this Hertz concluded that cathode rays could not possibly consist of charged particles but *had* to be some form of radiation. Then, in 1892, he showed that cathode rays could penetrate thin metal foils placed in their path. This observation reinforced his view that cathode rays could not consist of particles; while radiation could pass through solids, for example, light through glass, it seemed unlikely to Hertz that a solid could be transparent to particles. The scientific community now had a picture of cathode rays as some kind of radiation that could be deflected by a magnetic field — a new and rather strange concept.

7.3 J. J. Thomson and the Electron

In 1897 Joseph John Thomson (Figure 7.3; Nobel Prize for Physics, 1906), the Cavendish Professor of Physics at Cambridge University, eventually revealed the true nature of cathode rays. The critical experiment he carried out was to show that cathode rays *could* be deflected by electric fields and that Hertz's null result was due to the conditions of his experiment. The air pressure in Hertz's experiment was too high and consequently the effect of passing an electric current through it was to create in it what we now call a *plasma*, an intimate mixture of positively and negatively charged particles. The imposed electric field causes these charged particles to separate slightly, positive charges going one way and negative charges the other. This separation

Atoms Have Structure 59

Figure 7.3   J. J. Thomson (1856–1940).

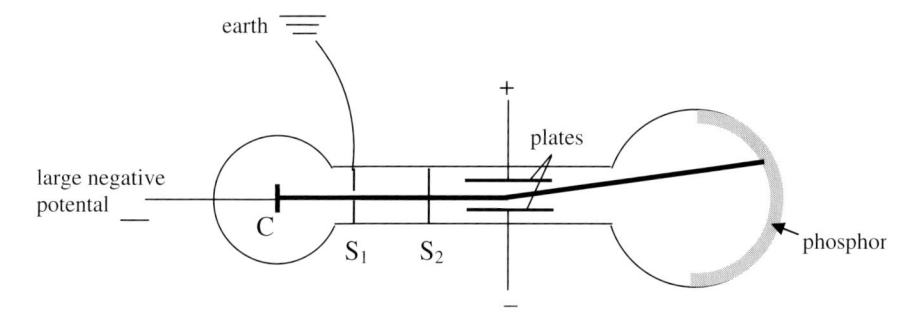

Figure 7.4   A schematic representation of Thomson's cathode ray experiment.

of positive and negative charges set up an electric field in the opposite direction to the imposed field. The movement of charges continued until the imposed field was neutralized within the equipment — at which stage the separation of charges ceased. This meant that there was no effective electric field within the equipment to deflect the cathode rays. Thomson used a much lower pressure in his apparatus, such that the density of charged particles in the plasma was too low greatly to affect the applied field.

The cathode-ray equipment used by Thomson is shown schematically in Figure 7.4. The cathode C is at a negative potential of several hundred volts relative to the anode S_1, which is earthed (zero potential). The two barriers S_1 (the anode) and S_2 contain narrow slits at right angles to each other so that a fine beam of cathode rays emerges from S$_2$. The metal plates have an electric potential difference between them so creating a field through which the cathode ray beam passes. The beam is deflected upwards towards the positive plate and is recorded by producing a bright patch on the phosphor-coated screen.

It was already known that cathode rays could also be deflected by a magnetic field. In order to get a deflection in an up and down direction the magnetic field must point in a direction perpendicular to the plane of the figure. A magnetic field was applied by a pair of coils through which a current was passed, one on each side of the apparatus, and the strength of the field was calculated from the dimensions of the coils and the current passing through them.

The aim of the Thomson experiment was to confirm that cathode rays consist of particles of some kind possessing both a charge e and a mass m. The force on a particle due to an electric field E depends on the product eE and the effect of that field in producing a deflection in its path depends on the mass of the particle. The more massive the particle the less it will be deflected. For a magnetic field B the force on the particle depends on the product eBv where v is the velocity of the particle perpendicular to the direction of the field. By adjusting the electric and magnetic fields so that they oppose each other to give zero deflection, the velocity of the cathode rays could be determined from the ratio E/B. Then by measuring the deflection from the electric field alone, and using the determined value of v, Thomson was able to estimate the quantity e/m, the ratio of the charge of the particle to its mass — but not the individual quantities forming that ratio. It was also found from the experiment that e is negative.

It was thought by many scientists who had worked on cathode rays that they moved with the speed of light, 300,000 kilometres per second, a belief that was based on the idea that they were a form of radiation. Thomson's result for velocity was less than one-tenth of the speed of light. We have seen that from electrolysis experiments the 'unit charge' had been determined and from Avogadro's constant the mass of a hydrogen atom, the lightest known atom, could be determined. Assuming that the charge on the cathode ray particle corresponded to the unit charge estimated from electrolysis, this gave a value of e/m for hydrogen, as estimated by Thomson, to be about 1,000 times less than the cathode ray result — meaning that the particle had about one-thousandth of the mass of a hydrogen atom (now we know that the ratio is $1/1\,837$).

Thomson called this new particle a 'corpuscle' and he announced its existence at the Royal Institution in London in April 1897. However, other scientists preferred the name 'electron', originally coined by Stoney, and that is the name by which it is now known. Thomson found that no matter what gas was used in his apparatus the value of e/m was always the same and he concluded that the electron was a constituent part of all matter regardless of what it was. The exact role that the electron played in the structure of matter was not known at the time. In an article in the *Philosophical Magazine* of October 1897 Thomson wrote 'We have in the cathode rays, matter in a new state in which the subdivision of matter is carried very much farther than in the ordinary gaseous state; a state in which all matter is of one and the same kind; this matter being the substance from which all the chemical elements are made up'. Taken literally this excerpt might be taken to mean that Thomson thought that all elements consisted of electrons and nothing else but we know that this is not what he actually meant. Thomson realised that in order to have atoms that were electrically neutral it was necessary to have something else with positive charge to balance the negative charge of his corpuscles. In 1897 he proposed the *plum-pudding model* of atoms in which the negative corpuscles were embedded in a sphere of positive charge, much as the raisins are distributed in a plum pudding (Figure 7.5). This model was widely accepted at the time it was made — although what it was that constituted the cloud of positive charge was not defined.

7.4 The Charge and Mass of the Electron

In 1897 the existence of the electron (for such we shall call it although Thomson clung to the term 'corpuscle' for twenty years) was well established and it was believed that the magnitude of its charge was the same as that of the unit charge derived from electrolysis. Assuming

62   *Materials, Matter and Particles*

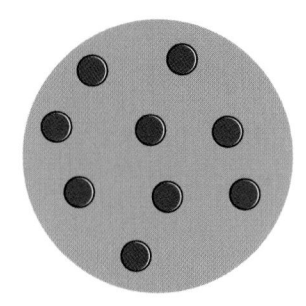

Figure 7.5   Thomson's plum-pudding model of an atom. Negatively charged corpuscles (electrons) are distributed within a spherical cloud of positive charge to give an electrically neutral atom.

Figure 7.6   Robert Millikan (1868–1953).

that is so then an estimate of the electron mass could be made from Thomson's value of e/m but clearly the better was the measurement of e the better would be the estimate of the electron mass. An ingenious experiment to determine an accurate value for e was made by the American physicist Robert Millikan (Figure 7.6: Nobel Prize for Physics, 1923) in 1909.

The apparatus used by Millikan is shown schematically in Figure 7.7. The enclosure shown has glass sides with the top and bottom as metal plates with a potential difference between them to give an electric field within the enclosure. Tiny oil drops are sprayed in through a nozzle in the top plate and the action of producing the spray induces

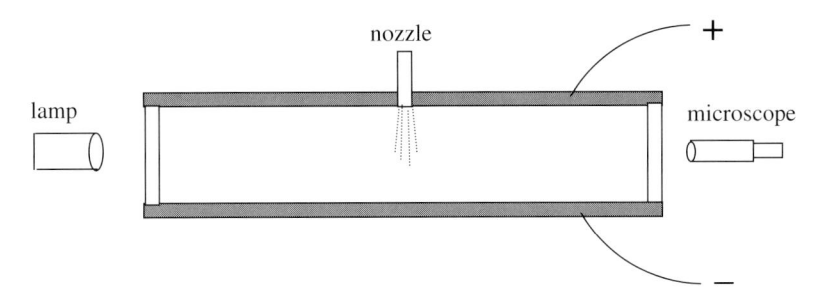

Figure 7.7 A schematic representation of Millikan's oil-drop apparatus.

charges on many of the drops. A lamp illuminated the drops, which were observed through a microscope.

The downward force of gravity on a drop is Mg, where M is the mass of the drop and g the acceleration due to gravity, which is known. The force due to the electric field is proportional to Eq, where E is the electric field, which can be accurately measured, and q is the total electric charge on the drop. If the electric field is adjusted so that a particular drop is stationary then the upward force on the drop due to the electrical force just balances that downwards due to gravity. If the mass of the drop were known then this would give the charge q on the drop. To estimate the mass of the drop Millikan switched off the electric field and observed the rate at which the drop fell. When a spherical object falls in a fluid — the air in this case — it eventually reaches a limiting speed, called the *terminal velocity*, which depends in a known way on the viscosity of the fluid, a measure of its resistance to flow, and the density and radius of the sphere. Since the viscosity of air is known, as was the density of the oil, the radius of the drop could be determined from its terminal velocity, as ascertained by observation through the microscope. Actually, for various technical reasons, Millikan's experiments involved measuring the terminal velocity both with the field switched off and with the field so high that the drop rose upwards. However, the essentials of the method are as described.

Millikan's results showed that charges on the oil drops were all small multiples of a small charge, which he found to be 1.592×10^{-19} C.

The interpretation of this is that the charges on the drops were due to them having picked up a few electrons in the process of being squirted from the nozzle. Thus they could have one electron charge, or two or more but that charge was strictly a quantized quantity. No drop could have a non-integral number of electron charges. This was an important result at the time. As well as giving a reasonable estimate of the electron charge (modern value $1.602\,176\,5 \times 10^{-19}$ C) it also showed that electric charge is not a continuously variable quantity. With the electron charge established then, from a good value of e/m, a good estimate of the mass of an electron could also be found; the modern value is $9.109\,382 \times 10^{-31}$ kilogram.

By the end of the first decade of the 20th century it was known that electrons were constituents of all atoms and that they contributed only a tiny proportion of the masses of the atoms. There was also the plum-pudding model of atoms but nobody had any idea of the nature of the pudding.

Chapter 8

Radioactivity and the Plum-Pudding Model

8.1 Röntgen and X-rays

The last decade of the 19th century and the first three decades of the 20th century was a period of momentous progress in the field of physics. Thomson's discovery of the electron in 1897 was the first step in elucidating the nature of atoms but many more steps were necessary before a complete picture would emerge. The first of these new steps can be traced back to 1895 when a quiet and modest German scientist, Wilhelm Conrad Röntgen (Figure 8.1) discovered the existence of X-rays.

The process of discovery began when Röntgen was investigating cathode rays using a vacuum tube in which there was a thin aluminium-foil window at the anode end that would allow the cathode rays to escape. It will be recalled that an experiment by Hertz showed that cathode rays could penetrate a thin metallic foil; this was the observation that had finally convinced him that cathode rays had to be some form of radiation. Because the aluminium foil was so flimsy and would be affected by electrostatic forces due to the strong field within the tube, Röntgen reinforced it with a thin piece of cardboard, which would still permit the passage of the cathode rays. Röntgen found that when another piece of cardboard coated with barium platinocyanide, a fluorescent material, was placed near the end

Figure 8.1 Wilhelm Conrad Röntgen (1845–1923).

of the tube it glowed, just as it would had it been exposed to visible light. Fluorescence is the phenomenon where a material becomes energized by absorbing radiation over a whole range of wavelengths but then re-emits the energy at one, or a few, wavelengths that are characteristic of the fluorescent material. In Röntgen's experiment the cardboard reinforcement for the aluminium foil prevented any light from escaping from the tube so it looked as though it was the cathode rays that had produced the fluorescence.

The vacuum tube used by Röntgen in this experiment was of a type with rather thin glass walls and he decided to repeat the experiment with a tube similar to those designed by Hittorf and Crookes, which had much thicker glass. In 1869 Hittorf had noticed a fluorescent glow in the glass at the anode end of the tubes he was using (§7.2). Röntgen covered the whole tube in black cardboard to ensure that no light of any sort could escape from it and to check that his cardboard enclosure was absolutely light tight he passed a current through the tube in a darkened room. Indeed, no light did come from the tube itself but Röntgen noticed a faint glow from a source some distance from the tube. He discovered that this source was a piece of cardboard covered with barium platinocyanide that he was intending to use in the next part of his experiment.

Röntgen came to the conclusion that some new form of radiation coming from the tube was responsible for the faint glow observed and he spent some time investigating its properties. He found that these rays, which he called *X-rays* to indicate their uncertain nature, would blacken a photographic plate. Experiments that involved measuring the blackening of a photographic plate by X-rays that had passed through materials showed that the transmission depended both on the type and thickness of the material through which they passed. Because of the differential absorption of bone and flesh he found that one could see an image of the bones of the hand by the transmission of X-rays recorded photographically. His first such radiograph, of rather poor quality, was taken of his wife's hand; a better radiograph of a colleague's hand is shown in Figure 8.2.

The most immediate application of X-rays was in medical radiography and to this day it remains an important diagnostic medical technique. When X-rays were first discovered the danger of overexposure to radiation was not known and X-ray generators were sometimes used as a form of entertainment in which, for fun, one would look at the bones within the body. There must have been many premature deaths due to this cause.

In the years to come, X-rays were to play an important role in the advancement of knowledge about the structure of individual atoms (Chapter 10) and also the detailed structure of matter as chemical combinations of atoms (Chapter 18) — something that has given us insights into the detailed processes of life itself (Chapter 20). These

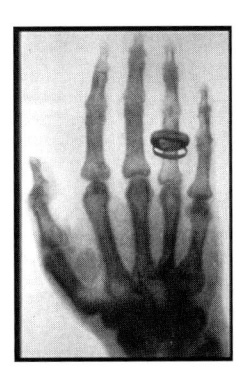

Figure 8.2 A radiograph taken by Röntgen of the hand of Albert von Kolliker.

fruits from the discovery of X-rays were still in the future but for the discovery of X-rays alone Röntgen deservedly won the very first Nobel Prize in Physics in 1901.

8.2 Becquerel and Emanations from Uranium

The next major advance in physics, following the discovery of X-rays, was made by the French physicist, Henri Becquerel (Figure 8.3). He came from a scientific family — his grandfather, father and son were all distinguished physicists — and his particular area of expertise was *phosphorescence*, a phenomenon related to luminescence. Fluorescence is the radiation emitted by a substance *while* it is being illuminated and phosphorescence is the radiation emitted by a substance *after* it has been illuminated. A phosphorescent material absorbs and stores energy from radiation falling on it and then releases that energy, at wavelengths characteristic of the material, over a period of time. Clocks and watches with so-called 'luminous dials' actually have phosphorescent dials!

In January 1896 Becquerel attended a meeting of the French Academy of Sciences at which Henri Poincaré (1854–1912), a French mathematician and theoretical physicist of extraordinary accomplishments in many fields, described Röntgen's recent work. In his talk he posed the question of whether the new X-radiation was connected in some way to the fluorescence in the glass at the end of the vacuum

Figure 8.3   Henri Becquerel (1852–1908).

tube, first noticed by Hittorf. With his background in a phenomenon similar to fluorescence, Becquerel saw this as an exciting area of investigation. If X-rays were simply an accompaniment to fluorescence then why not to phosphorescence, which involved similar light emissions with a delay built into to the process? Since he had many phosphorescent materials at his disposal the project was one that he could embark upon easily and with the minimum of delay. He began a series of experiments in which he wrapped a photographic plate in opaque black paper, placed the phosphorescent sample on top of it and then exposed the sample to sunlight so that it would emit light. If the very penetrating X-rays were also being produced by the sample then the photographic plate would be blackened. Becquerel tried one phosphorescent material after another but without success. Then, several weeks after he started his experiments, a sample of potassium uranyl sulphate that he had loaned to another laboratory was returned. At last he had success. The photographic plate was blackened and if he placed a coin or other opaque object under the sample then there was a corresponding clear portion of the plate imaging the object. In late February 1896 he announced his results at a meeting of the French Academy of Sciences and the linkage between phosphorescence and the production of X-rays seemed to have been established.

A few days after the Academy of Sciences meeting, Becquerel set up a few more experiments with crystals on top of wrapped photographic plates. However, since it was rather a dull day — too dull properly to expose his phosphorescent materials — he put them in a drawer until the weather improved. Some days later he removed the specimens from the drawer and, for reasons not absolutely clear, developed the photographic plates. The plate with which the potassium uranyl sulphate had been in contact was blackened — showing that phosphorescence was not needed to give whatever had exposed the plate! Becquerel realised that there was some kind of radiation being spontaneously emitted by the material and he announced his results to the French Academy of Sciences at its meeting on the following day.

Following his first announcement Becquerel carried out further experiments that showed that the radiation was emitted by any compound containing uranium, from which he concluded that it was the uranium atom itself that was the source.

8.3 The Curies and Radioactivity

The next major contributors in this area were a husband and wife team based in Paris. Maria Sklodowska was born in Warsaw but left in 1891 to join her sister in Paris because of political turmoil in Poland, at that time part of the Russian Empire. She entered the University of Paris where she studied chemistry, mathematics and physics. After a successful period of study, she graduated in 1893 with the best degree of any student of her year and one year later she obtained a master's degree. While at the university she married Pierre Curie (Figure 8.4a) and so became Marie Curie (Figure 8.4b), the name under which she was to achieve great international recognition and acclaim.

The Curies, working as collaborators, began a study of the subject that they were the first to call *radioactivity*. They decided to concentrate their investigations on pitchblende, the black mineral from which uranium is extracted. After extracting uranium from pitchblende the residue showed an even greater concentration of radioactivity than it had possessed originally. After many years of effort they extracted two

Figure 8.4(a) Pierre Curie (1859–1906).

Figure 8.4(b)   Marie Curie (1867–1934).

new radioactive elements from pitchblende — *polonium*, named in honour of Marie's native Poland, and *radium*, an intensely radioactive material that eventually had important medical applications, especially in the treatment of cancer. To her great credit Marie Curie refused to patent the process of extracting radium from pitchblende; she wished the benefits of her work to be available to everyone.

Becquerel and the Curies received the 1903 Nobel Prize for Physics for their pioneering work in radioactivity. So now the phenomenon of radioactivity was known to exist for more elements than just uranium — but what was the radiation?

8.4 Rutherford and the Nuclear Atom

Closely following the important and seminal French work on radioactivity, another star was beginning work in this area of science. This was Ernest Rutherford (Figure 8.5) a New Zealand physicist who, in 1898, was appointed as Professor of Physics at McGill University in Montreal, Canada. In 1899, while studying the ionization of gases (removal of electrons from atoms leaving them with a positive charge) by exposing them to the radiation from uranium and thorium, he discovered that there were two types of radiation, which he labelled α (alpha) and β (beta). He subjected the radiation coming from the radioactive atoms to a magnetic field and found that one part of it, the

Figure 8.5 Ernest Rutherford (1871–1937).

β radiation, was deflected while the rest, the α radiation, was unaffected. Using a similar experimental set-up to that used by Thomson a few years earlier (Figure 7.4) he measured the ratio e/m and showed that β radiation consisted of very energetic electrons — henceforth they could be described as β *particles*.

The two types of emission, α and β, differed greatly in their interactions with matter since α radiation was completely absorbed by a sheet of paper or a few centimetres of air while β particles could pass through a several millimetres of some metals. The following year, a French chemist Paul Villard (1860–1934) discovered yet another kind of emission from radioactive atoms, which he called γ (gamma) radiation. This carried no charge, rather like X-rays, but it had incredible penetrating power; it could pass through 20 centimetres of lead or a metre of concrete!

It eventually turned out that Rutherford's original experiment with a magnetic field had failed to bend the path of the α radiation because his field was too small. After many experiments, in 1902 he produced a field strong enough to deflect α radiation and showed that it had a positive charge and hence was almost certainly a stream of particles. When Becquerel heard of this result he confirmed it by independent experiments of his own.

In 1868 an English astronomer, Norman Lockyer (1836–1920), during a solar eclipse, was looking at some bright spectral lines emitted by prominences — huge streams of gas being ejected by the Sun. Laboratory experiments had shown that spectral lines are characteristic of particular elements, and most of the spectral lines he saw could be identified as those of known elements, but there was a very strong yellow line that corresponded to no known element. Lockyer eventually decided that it was due to some, as yet unknown, element and he called it *helium*, derived from the name of the Greek Sun god, *Helios*. Eventually helium was discovered by a Scottish chemist, William Ramsey (1852–1916), as a gas trapped within tiny cavities in the mineral *cleveite*, a radioactive mineral containing uranium. Helium was found to have an atomic weight of 4, smaller than that of any other element except hydrogen.

There were several lines of evidence emerging that suggested that α radiation consisted of particles corresponding to a helium atom that had lost two electrons and hence had a positive charge. The existence of helium trapped in cleveite suggested that it might be derived from the α particles emitted by the uranium in the mineral, which happened to pick up a couple of electrons and so become helium atoms. Again, by the use of very strong magnetic and electric fields, Rutherford had managed to measure e/m for α particles and so confirmed that they could correspond to helium atoms with a double positive charge acquired by the loss of two electrons. However the ratio e/m alone could not distinguish a doubly charged helium atom (four units of mass and two of charge) from a singly charged hydrogen molecule (two units of mass and one unit of charge). What the experiment did confirm is that what was being emitted was a particle and so the term α *particle* became firmly established.

In 1907 Rutherford moved to Manchester and when he arrived he set about the problem of confirming that the α particle *was* a doubly charged helium atom. It was possible to collect the α particles being emitted by a source in an enclosure and to measure the total charge on them. However, for this experiment to give the charge *per particle* is was necessary to count the rate at which the source was emitting the α particles. He was able to do this in 1908 through the work of a junior colleague, the German physicist Hans Geiger (1882–1945). Geiger had invented a device, the *Geiger counter*, which was capable of counting the number of α particles passing through it. When a particle entered a Geiger counter it ionized (knocked electrons off) some of the atoms of the gas within it. Just briefly, before the electrons recombined with the positively charged atoms to give a neutral atom again, the gas contained positive and negative charges and so was conducting. If an electrical potential was maintained across the counter then there was a brief discharge that could be translated into the click of a microphone and hence counted. Rutherford's measurements of the total charge and the number of particles giving that charge confirmed that there were about two positive units of charge per α particle.

All the indications were that the α particle was just a doubly charged helium atom but the final incontrovertible proof came from an experiment carried out in 1909. Rutherford asked the glassblower in his department to produce a vessel, the walls of which would be thin enough to permit the passage of fast-moving energetic α particles but thick enough not to allow the passage of helium atoms. This was done and α particles were fired into the vessel. After a few days the contents of the vessel were examined and found to give the characteristic yellow spectral line of helium. The α particles entering the vessel had picked up electrons and been converted into helium.

In 1908 Rutherford was awarded the Nobel Prize for Chemistry for his research into radioactivity but his greatest contribution to science was still to come.

While Rutherford was at McGill University he carried out experiments that showed that when α particles passed through matter they could be scattered. His first experiment, carried out early in 1906, used a wire coated with an α-emitter as a source with the α particles passing through a narrow slit parallel to the wire. When the experiment was carried out in a vacuum a photographic plate showed a sharp image of the slit. However, with air in the apparatus the image was blurred at the edges showing that scattering of the α particles by the atoms and molecules in the air had taken place. In a later experiment that year α particles were passed through mica and Rutherford noted that some of them had been scattered by as much as $2°$. That may not sound very much but we recall how difficult it was to deflect

α particles with magnetic and electric fields and Rutherford estimated that an electric field of 100 million volts per centimetre would be required to give such a deflection. He noted that the observation 'brings out clearly the fact that the atoms of matter must be the seat of very intense electrical forces.' As it turned out this was a very important observation.

In 1909 Geiger had under his supervision a young research student, Ernest Marsden (1889–1970) whom he was training as a researcher in radioactivity. One day Geiger mentioned to Rutherford that he thought that Marsden was ready to undertake a research project and Rutherford suggested that Marsden should look for possible reflections of α particles from a metallic surface. The experimental arrangement set up by Marsden to do this is shown in Figure 8.6. The radioactive sample was situated at S and the detector, D, was shielded from direct irradiation by the sample by the lead slab L. A thin gold foil, transparent to α particles was placed at F. Any α particles that were reflected by F towards D would be recorded. There was no great expectation that the experiment would give a positive result — what was expected was that all the α particles would pass straight through the foil, some of them with slight deviations — as for the mica experiment. At that time Rutherford believed in the plum-pudding model of the atom and such an intermingled arrangement of positive and negative charges would not give local electric fields that could do more than slightly affect the paths of the α particles.

The experiment was first run with the gold foil absent just to confirm that any detection of α particles would be due to the foil and not other parts of the apparatus. Then the gold foil was introduced.

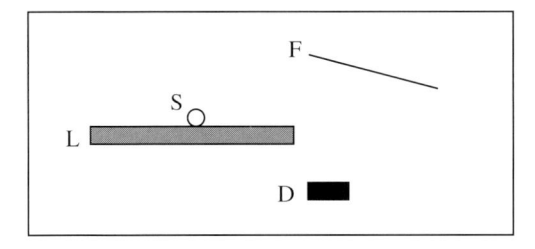

Figure 8.6 A schematic representation of Marsden's experimental arrangement.

To everyone's astonishment the detector immediately began to record the arrival of α particles. Rutherford likened the result to naval shells bouncing off a sheet of tissue paper!

Since an individual plum-pudding atom could only give a small deviation of an α particle the first thought was that the large deviations recorded were due to an accumulation of small deviations, all in the same direction, from a succession of α particle interactions. For the next year or so Rutherford, Geiger and others were trying, but unsuccessfully, to explain Marsden's results in terms of multiple scattering but no theory they came up with explained the facts. About 1 in every 8000 α particles was scattered backwards and no theory of multiple scattering could give anything close to that result. Then the light dawned — Rutherford hit on the solution — the plum-pudding model was wrong! A *single* scattering process from an isolated and tiny concentration of positive electric charge could explain the results. This would repel a positively charged α particle and any particle that just happened to approach very closely to the concentration of positive charge would be reversed in its tracks, or nearly so. The distance of approach required to give the back reflections of α particles was extremely small and it followed that the concentration of positive charge had to have a radius at least as small as that distance. A new model of the atom had emerged in which there was a tiny highly concentrated nucleus that contained all the positive charge of the atom and nearly all of its mass. Surrounding the nucleus, and at great distances compared with the size of the nucleus, were the electrons that made the atom neutral, gave the atom most of its volume but accounted for only a tiny part of its mass.

A mathematical analysis of the proportion of α particles that should be scattered at various angles from such a nucleus agreed with the experimental results and there was no room for doubt that Rutherford's nuclear model of an atom was substantially correct. It was a decisive step in understanding the nature of the atom. However there were more steps that still had to be made.

Chapter 9

Some Early 20th Century Physics

In the development of modern science, results from seemingly unconnected areas can impinge on the topic of interest — in our case the nature of matter. Here we give brief accounts of some important developments in the early part of the 20th century that, in one way or another, contributed to our understanding of the structure of atoms in particular and of matter in general.

9.1 The Birth of Quantum Physics

In 1900 the German physicist, Max Planck (Figure 9.1), at the University of Berlin, was trying to explain the intensity distribution with wavelength of the electromagnetic radiation from hot bodies. Electromagnetic radiation covers a wide range of wavelengths from radio waves of longest wavelength through to γ-rays of extremely short wavelength (Figure 9.2). It will be seen that the visible region is a very small part of the whole range.

The way that hot bodies radiate was well known from accurate experimental measurements and is, in crude measure, also a matter of everyday observation. A body at a temperature below a few hundred degrees centigrade, say an iron used for pressing clothes, emits heat radiation but no light. If a body, say a piece of iron, is gradually increased in temperature then first it will give off a dull red light, then the light will change to a yellow colour, then white and finally at extremely high temperatures the white light will be tinged with blue. What is happening is that as the temperature increases so the

78   *Materials, Matter and Particles*

Figure 9.1   Max Planck (1858–1947).

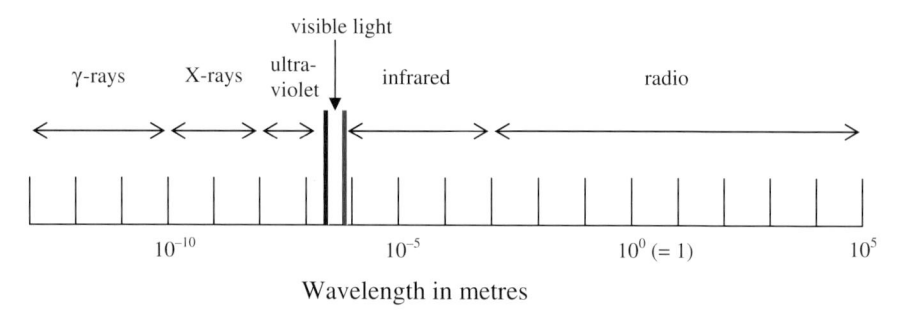

Figure 9.2   The electromagnetic spectrum. Each division gives an increase of wavelength by a factor of 10.

wavelength of greatest emission gets less — that is, goes from infrared towards ultraviolet. The variation of the intensity of emitted radiation with wavelength is shown in Figure 9.3 for three temperatures — 10 000 K, 12 500 K and 15 000 K — which shows the displacement of the peak to shorter wavelengths as the temperature increases. The figure also shows that the total amount of radiated energy, indicated by the areas under the curves, increases sharply as the temperature increases. When Planck applied the existing standard theory of radiation

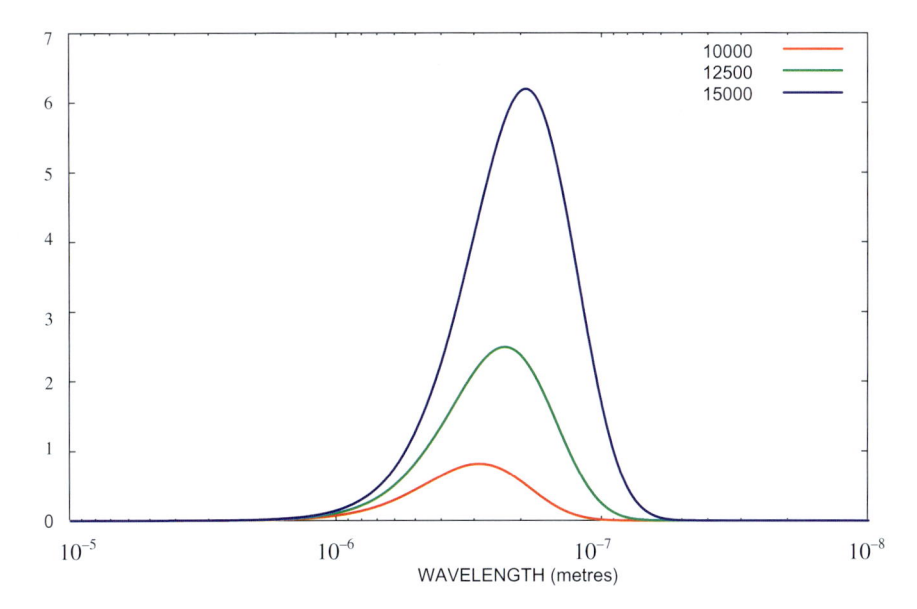

Figure 9.3 The variation of intensity of radiation with wavelength.

to try to explain these distributions he came up with a solution that was clearly nonsensical, predicting that at the ultraviolet end of the spectrum there would be an infinite amount of energy — a problem called the *ultraviolet catastrophe*.

After wrestling with this problem for a long time, trying one idea after another, Planck came up with a solution to the ultraviolet catastrophe that was radical and that would eventually revolutionize physics. The new feature that he put in to his analysis is best expressed in terms of the frequency of the radiation, v, the number of oscillations the wave makes per second, which is given by the speed of light divided by the wavelength. What Planck postulated was that electromagnetic radiation of frequency v was emitted by oscillating electric charges within a body but that an oscillator with a frequency v could only possess and emit energy in discrete units hv where h is a constant now known as *Planck's constant*. This means that the energy of an oscillator of frequency v can be hv, $2hv$, $3hv$ and so on and cannot not have any value in between these discrete energies. With this condition imposed the theory gave perfect agreement with observation. For this landmark discovery of what came to be known as *quantum physics* Max Planck was awarded the Nobel Prize in Physics in 1918.

9.2 The Photoelectric Effect

During the latter part of the 19th century, physicists observed that exposing a metallic plate to light could give interesting electrical effects. In 1899 J. J. Thomson shone light onto a cathode within a vacuum tube and observed that a current passed, indicating that electrons were being emitted. In 1902 an Austrian physicist Philipp von Lenard (1862–1947) made a detailed study of this phenomenon, which was named the *photoelectric effect*. His apparatus is shown in schematic form in Figure 9.4.

Von Lenard used a powerful arc lamp to illuminate the metal plate situated in a vacuum. He found the energy of the emitted electrons by finding the least possible potential difference between the metal plate, regarded as one electrode, and the other electrode at a negative potential with respect to the metal plate, which would just stop a current flowing. Alternatively, by having the other electrode at a positive potential with respect to the metal plate a current *would* flow through the tube, the magnitude of which would be a measure of the rate at which electrons were being emitted. His general conclusions were that the energy of the emitted electrons increased linearly with

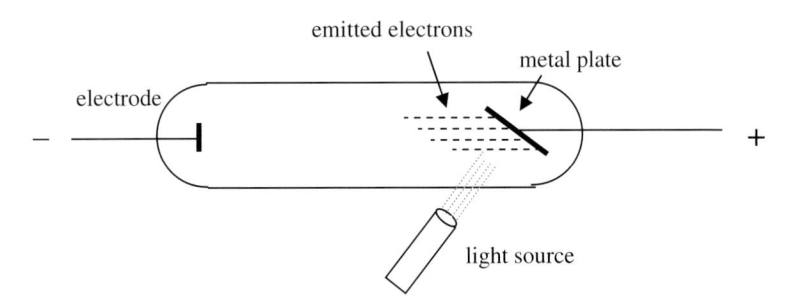

Figure 9.4 A schematic representation of von Lenard's apparatus.

Figure 9.5 Albert Einstein (1879–1955).

the frequency of the light used and that the rate at which they were emitted was proportional to the intensity of the light. If the frequency of the light were below a certain threshold then no matter how intense the beam of light no *photoelectrons*, as they came to be called, would be emitted. Von Lenard was awarded the Nobel Prize for Physics in 1905 for his work on cathode rays.

These results were explained in 1905 when Albert Einstein (Figure 9.5) wrote the paper that earned him the Nobel Prize for Physics in 1921. Einstein's explanation was in terms of Planck's quantum theory description of radiation. According to Einstein, light of frequency v existed in the form of packets of energy, called *photons*, each with energy hv. Interactions of the light with the metal plate consisted of individual photons hitting individual electrons attached to the metal atoms. It took a certain minimum energy to prise an electron away from an atom and then remove it from the plate, and unless the energy of a photon was greater than this minimum energy no photoelectrons would be produced. This explains the observed threshold frequency below which the photoelectric effect does not occur. Once above that frequency then the maximum energy acquired by the electron would increase linearly with the energy, and hence the frequency, of the photon. The rate at which photons struck the plate was directly proportional to the intensity of the light and, hence, so was the rate of emission of photoelectrons and also the measured current. This gave a perfect explanation of Von Lenard's observations.

9.3 Characteristic X-rays

Following the discovery of X-rays by Röntgen in 1895 they became the focus of a flurry of research attempting to elucidate their true nature. After a period of some uncertainty about whether they were particles or waves it was eventually demonstrated that they were a form of electromagnetic radiation with wavelengths of the order of one-thousandth of that of visible light. A characteristic of waves is that they can be diffracted — that is that they can bend around obstacles in their path and spread out after passing through a small aperture. If light is passed through a narrow slit then a *diffraction pattern* can be seen consisting of a set of light and dark fringes parallel to the slit. A diffraction pattern can also be seen when light passes through any kind of small aperture. Figure 9.6 shows the diffraction pattern when light is passed through a small square hole. The clinching experiment that X-rays were waves was carried out by two Dutch scientists, Hermann Haga (1852–1936) and Cornelius Wind (1861–1911) who passed X-rays through a very narrow slit and observed a fringe pattern around the image of the slit.

The X-rays that were first produced by Röntgen and his successors covered a wide range of wavelengths with a smooth variation of intensity throughout the range. The shorter the wavelength (higher the frequency) of X-rays the more penetrating they are and the distribution of intensity with wavelength could be determined by measuring the proportion of radiation passed through metal plates of varying thickness. However, an English physicist Charles Barkla (1877–1944)

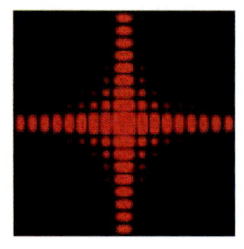

Figure 9.6 The diffraction pattern when light passes through a small square hole in a plate.

relative intensity

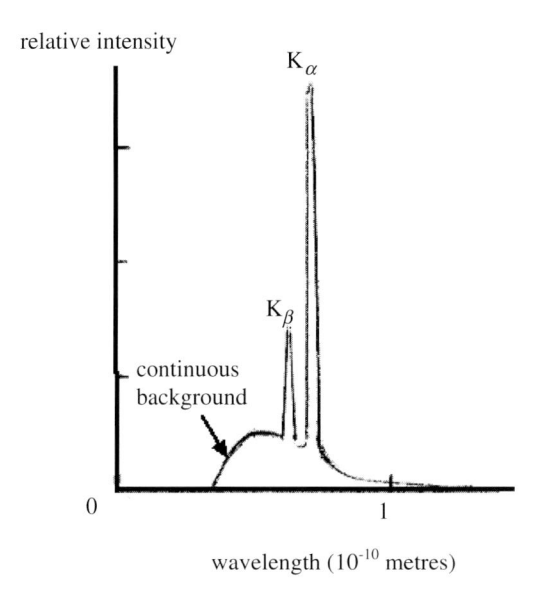

Figure 9.7 The characteristic X-radiation from molybdenum.

made an important discovery concerning the emission of X-rays by different elements. He discovered that, in addition to the general emission of X-rays covering a wide band of wavelengths, as described above, there was also the emission of sharp peaks of intensity at particular wavelengths. The wavelengths of this *characteristic X-radiation* depended only on the emitting element and were independent of its chemical association with other elements. However, in order to produce the characteristic radiation it was necessary to have a minimum potential difference between the anode and cathode, the magnitude of which varied with the material of which the anode was made. The form of emission is shown in Figure 9.7 for the element molybdenum. The two peaks, labelled K_α and K_β have wavelengths around 7×10^{-10} metres (70 nanometres) and stand out sharply against the continuous background. Barkla received the Nobel Prize for Physics in 1917 for his discovery of characteristic radiation.

It is possible by passing X-rays through metallic foils to filter out most of the X-radiation coming from a conventional source and just to transmit a very narrow band of wavelengths centred on one of the characteristic wavelengths. This provides a strong beam of monochromatic X-radiation that can be an important tool in many areas of science. In Chapter 18 we shall see how X-rays, mostly used in the form of monochromatic beams, have enabled the structure of matter to be examined at the atomic level.

Chapter 10

What is a Nucleus Made Of?

10.1 First Ideas on the Nature of the Nucleus

By 1911 it had been established that atoms contained negatively charged electrons of very small mass and a very compact nucleus that contained virtually all of the mass and sufficient positive charge to counterbalance the electron charges and so make the atom electrically neutral. This had been the contribution of physics towards an understanding of the nature of the atom. The chemists had also made a contribution in the form of the Periodic Table that incorporated knowledge of the chemical properties of the elements. The order of the elements in Mendeleev's Periodic Table, which was indicated as the *atomic number*, did not necessarily follow the order of the atomic masses. Sometimes, in order to preserve the chemical similarities within columns, on which the table depended, Mendeleev had to reverse the order of the atomic masses. Thus tellurium with atomic mass 127.6 came before iodine with atomic mass 126.9. Another pattern that clearly emerged is that for lighter elements the atomic mass tends to be twice the atomic number, or thereabouts, but as elements become heavier the ratio of atomic mass to atomic number becomes greater.

At this stage nobody had any idea of what constituted a nucleus; all that was known was that it was compact and had a positive charge. In a letter to the journal *Nature* in 1913 a Dutch scientist A van den Broek (1970–1926), discussing Rutherford's recent nuclear-atom discovery in the light of the Periodic Table, hypothesized that 'if all elements be arranged in order of increasing atomic number, the number of each element in that series must be equal to its intra-atomic charge'. This suggested that the atomic number, as given by the order in the Periodic Table, also gave the magnitude of the nuclear charge in electron units. Since atoms had to be electrically neutral this also gave the number of electrons associated with each atom. This was a revolutionary idea but still just a suggestion that needed to be tested in some way.

10.2 Moseley's Contribution to Understanding Atomic Numbers

A young British physicist, Henry Moseley (Figure 10.1), working in Rutherford's department in Manchester, now enters the scene to make an important contribution. He thought that it would be interesting to find the way that X-ray characteristic wavelengths (§9.3) varied with the element from which they came. Although Rutherford was not convinced that it was a useful thing to do, nevertheless he allowed Moseley to go ahead with the project.

The basis of Moseley's experiment was to set up what was essentially a cathode-ray tube but one in which the anode material could be changed from one element to another and which could be run

Figure 10.1 Henry Moseley (1887–1915).

with a high enough potential difference between the cathode and anode to excite characteristic X-radiation from the anode. The emitted X-radiation was then directed into an X-ray spectrometer, a device that made it possible to measure the intensity as a function of wavelength. The characteristic radiation could be identified by its high intensity (Figure 9.7) and so its wavelength could be determined. Having made his measurements for a wide range of anode materials Moseley then looked for a relationship that linked the characteristic wavelength to the atomic number in some systematic way. What he discovered is that if the square root of the frequency of the characteristic X-radiation was plotted against the atomic number one obtained a perfect straight line. The form of this relationship is shown in Figure 10.2.

These results gave a new authenticity to the concept of atomic numbers. They had been regarded just as an order, usually but not always following the order of atomic masses, which gave a chemically

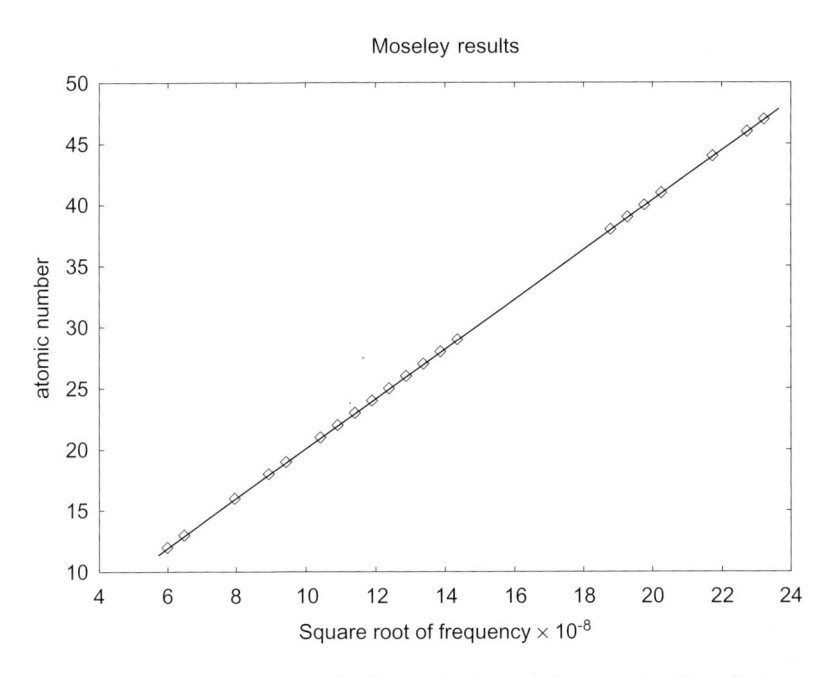

Figure 10.2 Moseley's results for a selection of elements for K_α radiation.

consistent Periodic Table. For example, to make sense of the Periodic Table cobalt and nickel had been given the atomic numbers 27 and 28, although their atomic masses indicated the opposite order. From Moseley's results it was clear that the order chosen on chemical grounds was the correct one and had some physical basis. From the results it could also be inferred that there were gaps in the atomic numbers corresponding to unknown elements — elements that were later discovered. The Periodic Table, as determined by Mendeleev, had atomic numbers going from 1, corresponding to hydrogen, to 92, corresponding to uranium. Of these 92 elements 4 do not occur naturally since they are so radioactive that they have not survived from the time of their formation — so there are only 88 naturally occurring elements.

A sad footnote to the Moseley story is that he died in 1915 in World War I in the Gallipoli campaign in Turkey; a promising career cut short.

Even before Moseley's work there was evidence that the hydrogen atom was, in some ways, a basic unit of atomic structure. In 1886 a German physicist, Eugen Goldstein (1850–1930), whose main research was on cathode rays, discovered that there was also radiation (actually particles but he did not know that) travelling in the opposite direction to the cathode rays in the vacuum tube. If channels are bored through the cathode then this radiation passes through the cathode; Goldstein called these rays *kanalstrahlen* or *canal rays*. Experiments in 1907 showed that canal rays could be deflected by magnetic and electric fields and hence were charged particles. Determinations of e/m for these particles showed that the lightest particles constituting canal rays were found when the tube contained hydrogen. Assuming that the positive charge had the same magnitude as that of the electron the particles were 1837 times as massive as the electron.

This evidence, coupled with the knowledge of the nuclear atom, the Periodic Table and Moseley's results, was leading to the general understanding that chemical reactions just involved the atomic electrons and that X-radiation was related to the charge on the nucleus in some way. The systematic variation of the wavelength of characteristic radiation with atomic number led to the belief that the atomic number represented the number of positive charges in the nucleus. The nucleus of hydrogen, the simplest and lightest element, had a single positive charge and the mass of the nucleus was the mass of that positively charged particle, a particle that Rutherford called the *proton*.

10.3 Breaking Up the Nucleus

In 1919 Rutherford carried out a remarkable experiment that, for all its simplicity, was the starting point in the branch of physics, now noted for its massive equipment and huge resource requirements, which came to be known as *nuclear physics*. Rutherford was carrying out experiments to measure the range of α particles emitted by metal sources coated with radioactive materials. The experimental arrangement is shown schematically in Figure 10.3.

The silver plate had a stopping power for α particles equivalent to about 6 centimetres of air, which is the range of the particles so that none of them should have left the box, despite the fact that the box had been evacuated. However, scintillations *were* seen on the screen and from the nature of the scintillations they were identified as 'fast hydrogen atoms', which we now know were hydrogen nuclei, or protons. They were recognized as such because if α particles are passed through hydrogen gas they collide with hydrogen atoms, imparting energy to them so that they become very energetic, that is, fast. The scintillations from such particles have a characteristic appearance and were readily recognized. Rutherford was not able to identify the source of the hydrogen atoms but thought that they probably came

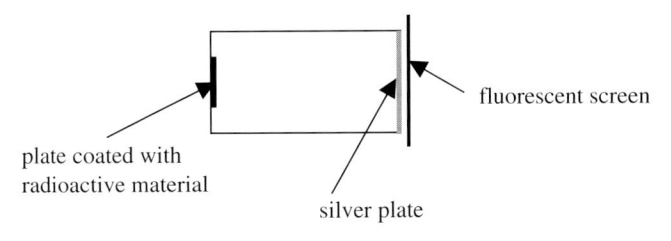

Figure 10.3 Rutherford's apparatus.

directly from the radioactive material. They were certainly not fast electrons — β particles — because these had been heavily deflected by a magnetic field and could not have reached the silver plate.

When the apparatus was filled with either oxygen or carbon dioxide the rate of scintillations fell, as would be expected if the protons came from the radioactive material, because of their absorption by the gas. The big surprise came when the apparatus was filled with air for then the scintillation rate *increased*! This was quite unexpected and since the effect was not observed with oxygen it was assumed that it was the major 80% component of air, nitrogen, which was causing the phenomenon. When the apparatus was filled with pure nitrogen the effect was enhanced by about 25% over the air value, which confirmed the view that nitrogen was the cause. The most obvious conclusion is that α particles were colliding with nitrogen atoms and that an energetic proton had been ejected from the nitrogen. We now know that a nuclear reaction that had taken place, which was

$$\text{nitrogen} + \alpha \text{ particle} \rightarrow \text{oxygen} + \text{hydrogen}$$

and that the 'fast hydrogen' recorded when the box was evacuated was due to the small amount of residual air in the box.

It is often stated that Rutherford had proved that the interaction of an α particle with nitrogen could give a proton and an oxygen atom but this is not strictly true. Rutherford did show that a nitrogen atom could be disintegrated to give a proton plus something else but he did not *prove* that the something else was oxygen. He did offer a hint that this might be so. When an α particle reacts with a nitrogen atom the heavy residue becomes, in Rutherford's words, '*fast nitrogen*'. He then notes that this fast nitrogen behaves like oxygen in its absorption characteristics but he never stated explicitly that oxygen was a product of the interaction. Nevertheless, Rutherford gave the first positive demonstration that one element could be changed to another. It was not transforming a base metal into gold but it was alchemy of a sort!

10.4 Another Particle is Found

Although by 1913 the hydrogen nucleus, or proton, had been identified as a constituent of all nuclei there remained the problem of reconciling the charge of the nucleus with its mass. For example, a carbon atom had a charge of six proton units but a mass of twelve proton units so the question arose of what was providing the extra mass. One explanation, superficially quite an attractive one, is that the nucleus contains protons and electrons, the latter partially neutralizing the charge of the protons. Thus a nucleus consisting of twelve protons and six electrons would give both the charge and the mass to explain carbon. However, one rather disturbing feature of the nucleus that was difficult to explain was how it could exist at all! The positively charged protons within a nucleus should be repelling each other so that the whole nucleus should explode outwards. To explain the stability of the nucleus Rutherford postulated that the extra mass in the nucleus, over and above that of the protons, was provided by neutral particles that acted as a kind of nuclear glue that prevented the protons from springing apart. The idea he expressed is that some intimate combination of a proton and an electron would effectively form a new particle with rather novel properties. The origin of the term *neutron* to describe this particle is not certain but may be credited to William Draper Harkins (1873–1951) an American nuclear chemist who postulated the existence of the neutron at about the same time as Rutherford.

We have seen that identifying the nature of charged particles — the electron and the proton — was comparatively straightforward because they were deflected by magnetic and electric fields and so the ratio of e/m could be determined. Not only did this give the ratio of charge over mass but also indicated clearly that what was being dealt with was actually a particle. The detection of a particle without charge would obviously be a much more difficult exercise.

The first relevant experiment was carried out in 1930 by the German physicist Walter Bothe (1891–1957; Nobel Prize for Physics, 1954) who used polonium as an α particle source to bombard the light element beryllium — atomic number 4, atomic mass 9.

He found that the beryllium gave off a very penetrating but electrically neutral radiation, which he interpreted as very energetic γ rays of the type given off by some radioactive elements. Then, in 1932, a daughter of Marie Curie, Irene Juliot-Curie (1891–1957) and her husband Frédéric Juliot-Curie (1900–1958), who jointly won the Nobel Prize for Chemistry in 1935 for their discovery of artificial radioactivity, decided to investigate the nature of the penetrating rays from Bothe's experiment. They directed the radiation at a slab of paraffin, a substance with a large component of hydrogen in its composition, and the result was the expulsion of very energetic protons from the paraffin. The analogy that seemed appropriate to them was the photoelectric effect where electrons within a metal surface are knocked out by the action of a photon of visible or ultraviolet light. In their interpretation of their experimental results the γ-ray photons coming from the beryllium had enough energy to eject protons (hydrogen nuclei) out of paraffin — despite the fact that a proton has 1 837 times the mass of an electron.

When the results of the Juliot-Curie experiment became known in Cambridge, to which university Rutherford had moved by this time, a certain amount of disbelief was expressed and James Chadwick (Figure 10.4), who was at the Cavendish Laboratory, decided to check the results and explore further. Chadwick had been interested in the idea of the neutron for some time but had been unable to find any evidence for its existence. Chadwick was convinced that the Juliot-Curie explanation of the expulsion of the protons by γ rays was wrong. The energy of the protons expelled from the paraffin, as assessed by deflection experiments, was about 50 million electron volts.[d] The energy of the α particles coming from polonium are 14 million eV and assuming that an α particle striking a beryllium atom produced a γ ray photon then the photon's maximum energy would be 14 million eV. It would clearly be impossible for such a γ ray photon to give more than three times its energy to a proton.

---

[d] The electron volt (eV) is the natural, but tiny, unit of energy in dealing with atoms and nuclei and is the energy acquired by an electron in moving through a potential difference of 1 volt. To put it in perspective the energy required to provide one kilowatt for one second is 6.25×10^{21} eV.

Figure 10.4 James Chadwick (1891–1974).

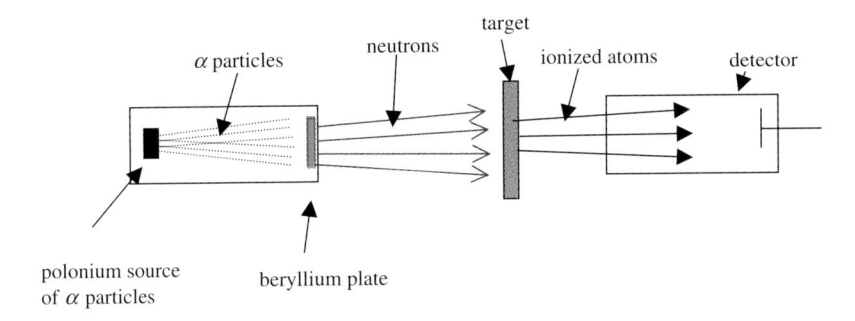

Figure 10.5 Chadwick's apparatus for investigating the neutron.

Chadwick was convinced that the penetrating 'rays' were actually neutrons and the experiment he set up to find their properties is shown in Figure 10.5. The α particles from the polonium source strike the beryllium plate and the 'penetrating rays' go on to strike the target, which could be a plate of paraffin wax or some other material, which could be a gas in a thin-walled vessel. The rays knock atoms out of the target and the collision process also ionizes these atoms by removing an electron. These charged particles, of known mass, could be counted by means of a Geiger counter and the energy they contained could also be measured by the total amount of ionization they produced in the gas of the detector. From the number of particles and their total energy he was able to determine the speed with which the ejected ionized atoms were moving. By repeating this experiment with various targets, getting different speeds for the ejected ionized atoms each time, and by applying the laws of mechanics to all the data for the different ejected atoms he was able to estimate both the mass and the velocity of the neutrons. The new particle was found to have a mass that was 1.0067 times that of a proton. For this work in establishing the existence of the neutron Chadwick was awarded the Nobel Prize for Physics in 1935.

Even after Chadwick's discovery of the neutron it was not *completely* certain that it was a new particle, completely distinct from being just a very intimate combination of a proton and an electron. However, within a couple of years it was established beyond reasonable doubt that the neutron was, indeed, a new particle. The German physicist, Werner Heisenberg (1901–1976) produced a theory that explained how an attractive force between a neutron and a proton could occur and how this attractive force could bind the nucleus together, despite the fact that it contained many particles all with a positive charge.

The structure of atoms seemed now to be established. Each consists of a compact nucleus containing protons and neutrons. The number of protons gives the atomic number of the element and the combined number of protons and neutrons gives the atomic mass in what are known as *atomic mass units* (amu). A neutral atom has a number of electrons, equal to the number of protons, which surround the nucleus and give the atom its finite volume. If the atom is bombarded, or energized by some other process, then one or more of the electrons can be removed from an atom to give an *ion* with a positive charge. The chemical properties of an element depend on how its electrons interact with the electrons of other elements. Radioactivity is produced by the breakup of the nucleus in some way with the loss of a particle or radiation and its transformation from one element to another.

The simplicity of the original Greek model of matter as consisting of combinations of four basic elements had been complicated by chemists who increased the number of elements to 92. Now physicists had restored simplicity again. There were only three basic ingredients to form all matter — protons, neutrons and electrons. However, we shall see that this simple picture did not last for too long.

Chapter 11

Electrons in Atoms

11.1 The Bohr Atom

Even before the existence of the neutron was confirmed it was known that atomic electrons existed in the space around a compact positively charged nucleus and there was considerable interest in knowing how they were distributed. An important step in understanding the electronic structure of atoms was made by the Danish physicist, Neils Bohr (Figure 11.1). After receiving his doctorate at the University of Copenhagen in 1911, Bohr went to England where he worked first with J. J. Thomson in Cambridge and then with Rutherford in Manchester. In 1913 he proposed a model for atomic structure in which electrons orbited the nucleus much as planets orbit the Sun except that, in place of the attractive force of gravity there was the attractive electrostatic force between the positively charged nucleus and the negatively charged electrons. However, unlike planets that, in principle, could orbit at any distance from the Sun, the distances of the electrons from the nucleus were constrained in Bohr's model.

The constraint that Bohr's electron orbits had to satisfy involved a physical quantity called *angular momentum*. For a body in a circular orbit this quantity is the product of the mass of the orbiting body, M, the radius of its orbit, r, and speed in its orbit, v. What Bohr postulated is that the angular momentum was quantized in units of $h/2\pi$ ($=\hbar$), where h is Planck's constant (§9.1) and π is the ratio of the circumference of a circle to its diameter. This condition, that the angular momentum had to be of the form $n\hbar$ where n is an integer, also

Figure 11.1 Neils Bohr (1885–1962).

restricted the energies of the orbiting electrons. The total energy of an orbiting electron is in two parts. The first part is the energy associated with the separation of the electric charges of the electron itself and the nucleus. This energy is negative in sign and is the *potential energy*, the magnitude of which indicates how strongly the electron is bound to the nucleus. The other component of the total energy of the electron is its energy of motion, *kinetic energy*, and the magnitude of this energy, which is positive in sign, indicates the tendency of the electron to escape from the nucleus. For an electron to be retained by the atom its potential energy must be dominant so that its total energy, which is the sum of the potential and kinetic energies, must be negative. The more negative the total energy is, the more strongly is the electron bound to the nucleus.

From the angular momentum constraint imposed by Bohr it can be found, using the equations of mechanics, that the possible energies that could be taken up by electrons in orbit around the nucleus vary as $1/n^2$ and, since the energies are negative, $n = 1$ gives the lowest energy (largest magnitude but negative) and hence gives the most strongly bound electrons. This is called the *ground state* of the electron. For $n = 2$, the energy is one-quarter of that in the ground state and for $n = 3$ it is one-ninth. As n increases the electron is less and less tightly bound.

Electrons in Atoms

In the foregoing discussion we have ignored a contribution to the potential energy of a particular atomic electron due to the presence of the other electrons. However, for hydrogen, which only has one electron, this complication does not arise so we shall restrict the following discussion to hydrogen. In Figure 11.2(a) we show the possible orbits that could be taken up by the electron in following Bohr's constraints on the angular momentum. The closest orbit is for $n = 1$ and as n increases so does the distance of the electron from the nucleus. In Figure 11.2(b) the energies corresponding to the different values of n are shown. They are all negative and as n increases so do the energies (that is they become smaller in magnitude and hence less negative).

When hydrogen is at a high temperature then, while most atomic electrons are in the ground state ($n = 1$), some are in higher energy states. All physical systems have a tendency to go towards states of lower energy so from time to time some electrons jump down from a higher to a lower energy state — for example, from $n = 2$ to $n = 1$ or from $n = 4$ to $n = 2$. When that happens then, according to the Bohr model, the released energy appears as a photon, a package of light energy as envisaged by Einstein in explaining the photoelectric effect. Thus, if the electron jumped from $n = 2$, with energy E_2, to $n = 1$,

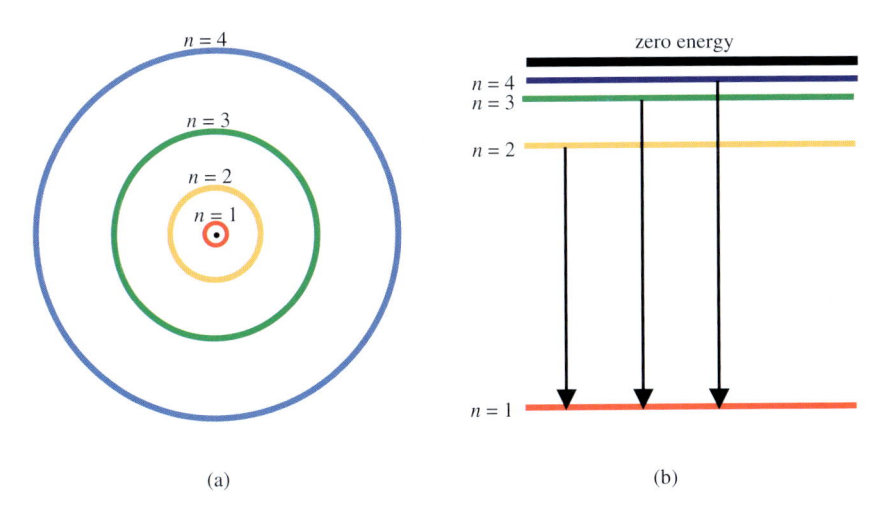

Figure 11.2 A Bohr model atom. (a) Electron orbits (b) Electron energies.

with energy E_1, then a photon with frequency v would be produced where $hv = E_2 - E_1$. In the heated hydrogen gas, atoms would be colliding with each other and in the collisions some electrons would take up energy and be nudged from lower to higher energy states. In this way the heated hydrogen would continuously emit light of specific frequencies (or wavelengths) and so give its characteristic spectrum.

In 1885 the Swiss mathematician Johann Balmer (1825–1898) noticed that within the hydrogen spectrum there was a set of lines with frequencies of the form

$$v = R\left(\frac{1}{2^2} - \frac{1}{n^2}\right)$$

for different integers n (greater than 2), where R is a constant. This set of spectral frequencies, all in the visible part of the spectrum, is known as the Balmer series. Then, in 1906 an American scientist, Theodore Lyman (1833–1897) found a similar series, but this time in the ultraviolet part of the spectrum, of the form

$$v = R\left(\frac{1}{1^2} - \frac{1}{n^2}\right),$$

now known as the Lyman series, with the same constant R, known as the Rydberg constant. The Bohr model for hydrogen explains these series and the value for R from the theory exactly matches that found by experimental measurements. The Lyman series corresponds to electron transitions to the ground state shown in Figure 11.2(b) by the vertical arrows. The Balmer series then corresponds to transitions from higher energy states to the state with $n = 2$. Other series of lines are known corresponding to transitions to $n = 3$, $n = 4$ and so on.

This perfect explanation of the hydrogen spectrum was a major triumph for the Bohr theory but, unfortunately, it did not work for more complicated atoms — although for some atoms like sodium and potassium it did have approximate validity.

11.2 Waves and Particles

Max Planck's explanation for the form of radiation curves in terms of the quantization of radiation, published in 1901, was the beginning of *quantum physics*, which profoundly changed our understanding of the physical world. Charged oscillators in a hot body could only have energies $nh\nu$ where n is an integer, h is Planck's constant and ν is the frequency of the oscillator. Planck's constant then appeared again in 1905 when Albert Einstein explained the photoelectric effect in terms of packets of radiation, called photons, all with energy $h\nu$ and they behave just like material particles of that energy. It is as though the electrons have been ejected from the metal surface by bullets of light. Radiation, characterized as a wave motion and showing this by its ability to be diffracted, could sometimes behave like material particles.

In 1924 a French research student, Louis de Broglie (Figure 11.3) produced a remarkable doctorate thesis in which he postulated that, just as waves can sometimes manifest the properties of material particles, so material particles can sometimes manifest the properties of waves. He gave an equation linking the two kinds of behaviour involving the *momentum* of a particle — the product of its mass and

Figure 11.3   Louis de Broglie (1892–1987).

its velocity — and the wavelength it would display when behaving as a wave. This relationship was of the form

$$\text{wavelength} = \frac{h}{\text{momentum}},$$

once again bringing in Planck's constant. The relationship, known as the *de Broglie hypothesis*, was simply that — a hypothesis — when it was first introduced.

An experiment carried out by two American physicists, Clinton Davisson (1881–1958) and Lester Germer (1896–1971) in 1927 confirmed the de Broglie hypothesis. Davisson predicted, on the basis of the de Broglie hypothesis, that it should be possible to carry out an experiment in which electrons are diffracted from a crystal,[e] the momentum of the electrons being adjusted to give a convenient value for the de Broglie wavelength. The Davisson and Germer experimental arrangement is illustrated in Figure 11.4.

Electrons were emitted by the hot cathode and accelerated towards the anode. A fine beam of electrons passed through a hole in the anode and fell on to a nickel crystal. In the actual experiment the potential difference between the cathode and anode was 54 volts that, according to the de Broglie hypothesis, should give electrons of

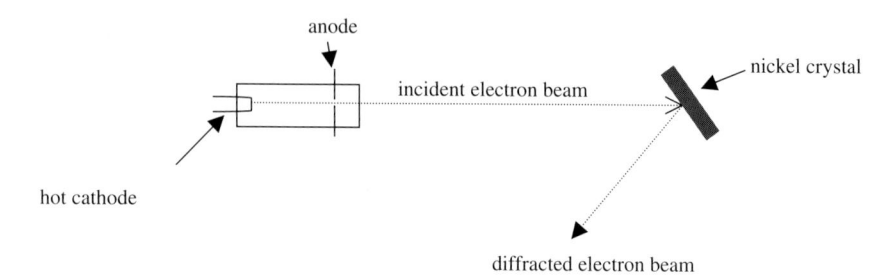

Figure 11.4 The Davisson–Germer experiment.

---

[e] The topic of diffraction of X-rays from crystals is dealt with in Chapter 18.

equivalent wavelength 16.7 nanometres (nm).[f] The crystal was set to diffract radiation of that wavelength with the diffracted beam at 50° to the incident beam. A distinct diffracted beam of electrons at this angle was detected. This was the very first experiment that demonstrated that entities normally thought of as particles — electrons — could have wave-like properties. An English physicist, George Paget Thomson (1892–1975), the son of J. J. Thomson, had also predicted the phenomenon of electron diffraction and he and Davisson were jointly awarded the Nobel Prize for Physics in 1937. The confirmation of his hypothesis also led to the award of the Nobel Prize for Physics to de Broglie in 1929, the first time the prize had been awarded for work contained in a doctorate thesis.

11.3 The Bohr Theory and Waves

There was growing evidence for what was becoming known as *wave-particle duality*, which is to say that matter could sometimes behave like waves and radiation could sometimes behave like particles. The rule seemed to be that if an experiment was set up to detect particle-like behaviour then that is what was observed and, likewise, an experiment designed to measure wave-like properties would measure just that. A single experiment might show both forms of behaviour. For example, in the Davisson–Germer experiment a beam of electrons was fired at a crystal and obediently gave a diffracted beam in the appropriate direction. The electrons were given the circumstances to behave like waves and they did so. However, when it was required to observe the distribution of diffracted electrons the detector used was one that treated electrons as particles; what was measured as the equivalent of the intensity of the diffracted beam was the frequency of arrival of individual electrons.

In his empirically derived model of the possible electron orbits in an atom, which certainly worked well for hydrogen, Bohr had postulated that the allowed values of angular momentum had to be integral multiples of $\hbar$. However, the electrons in their orbits not only had

---

[f] 1 nm is 10^{-9} m.

angular momentum, a property of their rotational motion, but also linear momentum, defined as the product of mass and velocity. A little elementary mathematics shows that the de Broglie wavelength derived from their momentum is simply related to the circumference of the Bohr orbit. From the Bohr condition for an allowed orbit

$$Mrv = nh = \frac{nh}{2\pi}.$$

From the de Broglie hypothesis

$$Mv = \frac{h}{\lambda}$$

and combining these relationships gives

$$n\lambda = 2\pi r,$$

which indicates that n de Broglie waves exactly fit into the circumference of the orbit ($2\pi r$). This relationship is illustrated in Figure 11.5 for the energy levels $n = 1, 2, 3$ and 4. The full-line circle represents the Bohr orbit and the dotted line the whole number of waves fitting into the circumference of the orbit.

This gives a requirement for the Bohr orbits expressed in a different form from that given by Bohr, which was chosen just to get the right energies to explain the spectrum of hydrogen and which gave an expression for the Rydberg constant. Now we can start with the de Broglie hypothesis, confirmed by the Davisson and Germer experiment, and say that the orbits are restricted to those for which the momentum of an electron is such that a whole number of de Broglie waves will fit around the circumference.

This curious schizoid behaviour of particles and radiation, which change their behaviour according to what is expected of them, had become clear by the middle of the third decade of the 20th century. Traditionally, the behaviour of electromagnetic radiation was described by the classical theory put forward by the Scottish mathematician and

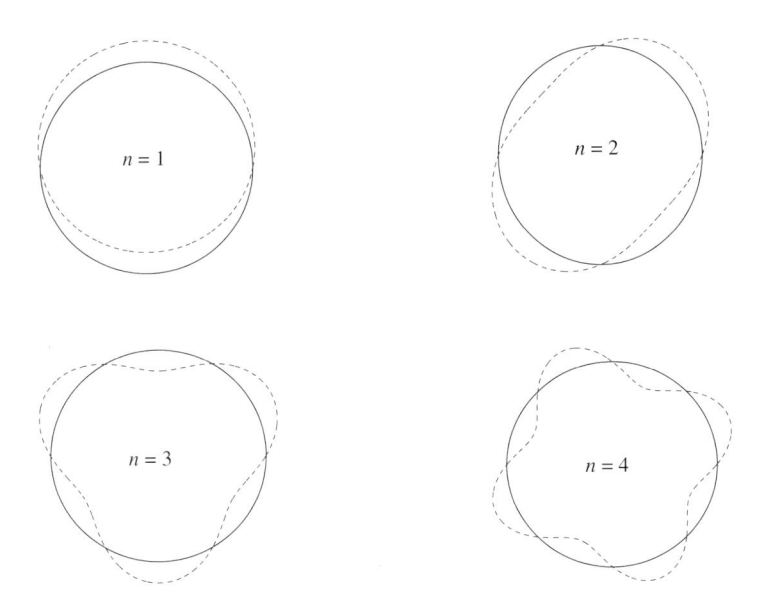

Figure 11.5 De Broglie waves fitted around Bohr orbits.

physicist James Clerk Maxwell (1831–1879) while the behaviour of particles was governed by the classical mechanics of Isaac Newton and the non-classical mechanics given by Albert Einstein in his Special and General theories of Relativity. How to reconcile these theories, which treated particles and waves quite separately, with the reality of modern observations, was exercising some of the best scientific minds and in the 1920s a theory that would do this and bring about a new revolution in physics was being developed in Germany (Chapter 12).

11.4 An Improvement of the Bohr Theory

When the spectrum of hydrogen was examined with greater precision it was found that the lines corresponding to the Bohr theory were not single lines but clusters of very closely spaced lines. The German scientist Arnold Sommerfeld (Figure 11.6) offered an explanation for this in 1922. The Bohr orbits, apart from the quantization of angular momentum, were defined by classical mechanics in just the same way as are the orbits of planets. However, planetary orbits are mostly elliptical, to a greater or lesser extent, and Sommerfeld proposed that

Figure 11.6 Arnold Sommerfeld (1868–1951).

electron orbits could also be elliptical. To explain the observed spectrum he postulated that for any value of n, which we now call the *principal quantum number* there were n possible orbits designated by n integers, l, called the *azimuthal quantum numbers* that could vary in value from 0 to $n-1$.

An impression of the possible orbits for $n = 1$ to 4 are shown in Figure 11.7.

All the orbits with the same value of n had very similar energy values but energies did vary slightly with the value of l. Thus an electron jumping from the orbit $n = 2$, $l = 1$ to the ground state ($n = 1$, $l = 0$) would give up a slightly different energy from one jumping from $n = 2$, $l = 0$ to the ground state. Hence the frequencies of the emitted photon would be slightly different and this would be seen as finely spaced lines in the spectrum.

Another improvement in obtaining agreement with the observed spectra came about by taking into account that the orbits could be in different planes. The best way of thinking about this is to imagine that the atom is placed in a magnetic field, which defines a direction in space. Now, a charged electron moving round an orbit is equivalent to an electric current in a loop circuit and such a current generates a magnetic field just like a magnet. For this reason the orbiting electron behaves just like a small magnet, the strength of which is proportional

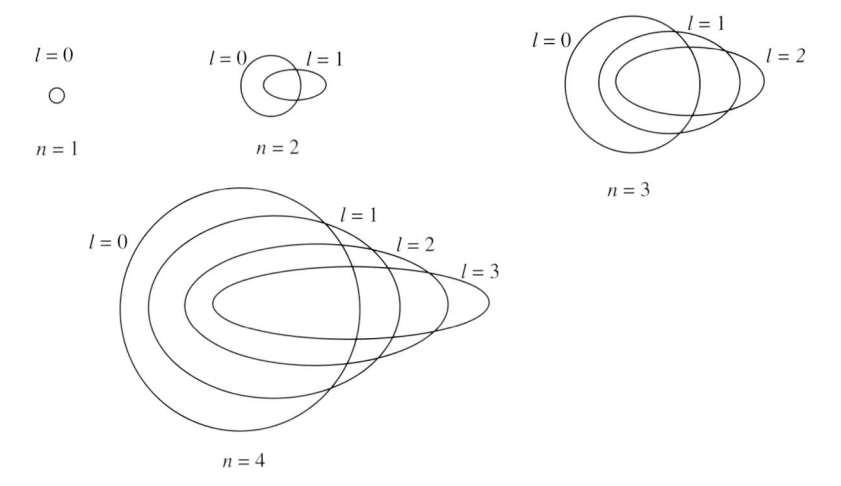

Figure 11.7 A representation of orbits for different values of n and l.

to the angular momentum in the orbit (Figure 11.8(a)). In classical physics when a magnet is placed in a magnetic field it aligns itself with the field, just as a compass needle always points towards magnetic north. Sommerfeld added the principle of quantization to the orientation of the orbit such that the component of the angular momentum along the direction of the magnetic field was mh, where m, the *magnetic quantum number* could be any integer between $-l$ and $+l$ — that is it could have $2l + 1$ different values. Figure 11.8(b) shows the possible orientations of an equivalent magnet for $l = 2$ according to the Sommerfeld model.

The value of the energy in an orbit was primarily due to n and, ignoring the closely spaced lines, these values explained the main series of lines — the Lyman series, Balmer series and so on. The introduction of l explained the closely spaced lines. If the atom were located in a magnetic field then the energies associated with the orbits would have a small dependence on their orientation. Consequently a further splitting of spectral lines occurred that could be explained as due to different values of m. This model worked very well for hydrogen but not for other atoms. Sets of quantum numbers that can occur are listed in Table 11.1.

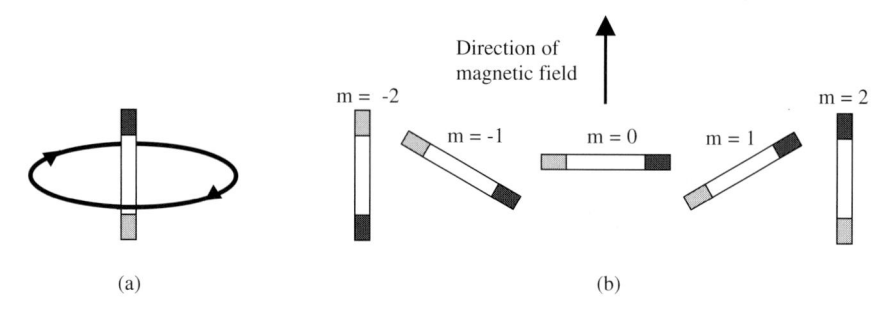

Figure 11.8 (a) The equivalent magnet for an electron orbit. (b) The possible orientations of the equivalent magnet (electron orbit) in a magnetic field for $l = 2$.

Table 11.1 Quantum numbers for $n = 1, 2, 3$ and 4.

n	l	m
1	0	0
2	0	0
	1	−1 0 +1
3	0	0
	1	−1 0 +1
	2	−2 −1 0 +1 +2
4	0	0
	1	−1 0 +1
	2	−2 −1 0 +1 +2
	3	−3 −2 −1 0 +1 +2 +3

This description of the hydrogen spectrum in terms of the three quantum numbers n, l and m was good, but not perfect. When spectral lines were examined at extremely high resolution they were found to consist of two lines with very tiny spacing. Several years were to pass before an explanation of this line splitting came available.

Chapter 12

The New Mechanics

12.1 Schrödinger's Wave Equation

During the period of the 20th century leading up to 1926 it had become clear that the simple view that there were two distinct areas of physics with separate theoretical approaches — those dealing with radiation and with particles — was no longer tenable. Entities such as light and electrons could appear either with wave-like properties or with particle-like properties, depending on the circumstances of the observation. It was clearly unsatisfactory that when electrons behaved like particles they were described by Newtonian mechanics that explained how they moved, whereas when they behaved like waves they were described by a wave equation that explained how they were diffracted. How much better it would be if a theory could be found that covered all aspects of the behaviour of electrons! That was the challenge taken up by the German theoretical physicist Erwin Schrödinger (Figure 12.1), the answer to which earned him the Nobel Prize for Physics in 1933.

To understand the basis of Schrödinger's approach we first need to understand the nature of wave motion. Consider a water wave moving along a canal that is being observed at two different places by Grey and Black (Figure 12.2). At Grey's position the height of the wave depends on time and the same is true for Black, but because they are at different positions they do not see the same height of wave at the same time. In general, the height of the wave as seen by any

Figure 12.1 Erwin Schrödinger (1887–1961).

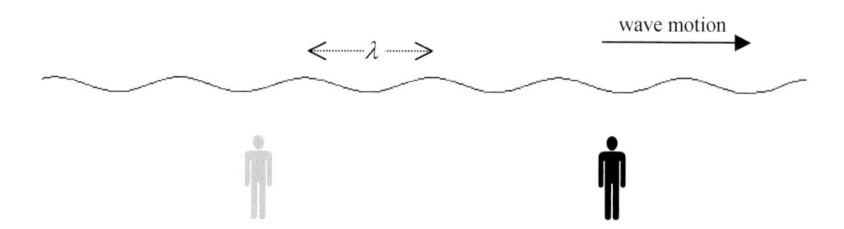

Figure 12.2 Waves on a canal as seen by different observers.

observer will depend on both the position and the time that the observation is made. This relationship can be written in a form of a *wave equation* in which the height of the wave is expressed in terms of position and time, an equation that also involves the wavelength of the wave, λ, and its frequency, v.

We have already seen that, in the relationship between waves and particles, the de Broglie wavelength has an equivalent momentum (h/λ) and in Planck's theory — for example, as applied to a photon — frequency has an equivalent energy (hv). Substituting equivalent momentum for wavelength and equivalent energy for frequency in the classical wave equation gives a new form of wave equation that involves momentum and energy, properties that are possessed by particles. Schrödinger postulated that this equation could be applied to the behaviour of a particle. This behaviour would be described by a wave-like quantity associated with the particle, equivalent to the height of the wave in Figure 12.2 and conventionally referred to as the *wave function*, represented by the symbol Ψ. The equation then gave the variation of Ψ with respect to position and time in terms of the momentum and energy of the particle. This is the Schrödinger wave equation (SWE) the origin of which is represented in Figure 12.3.

For any particular position the value of Ψ will oscillate around zero going between peaks and troughs in a periodic way. If the forces operating within the system do not change with time then the maximum height, or *amplitude*, represented by ψ, at a particular position does not depend on time. The way that ψ varies with position in such a situation can be derived from the SWE to give what is known as the time-independent Schrödinger wave equation (TISWE), which is all that is needed to describe many physical systems. In what follows we shall just be concerned with a general description of what wave mechanics tells us about various kinds of system and consider only cases where the TISWE is applicable.

12.2 The Wave Equation and Intuition

A feature of the SWE is that, in many applications, it only gives solutions for ψ for particular values of the energy of the particle. We take as an example the case of a particle undergoing a *linear harmonic oscillation*. This is an interesting example because you will recall that the assumption made by Planck was that a radiating body contained

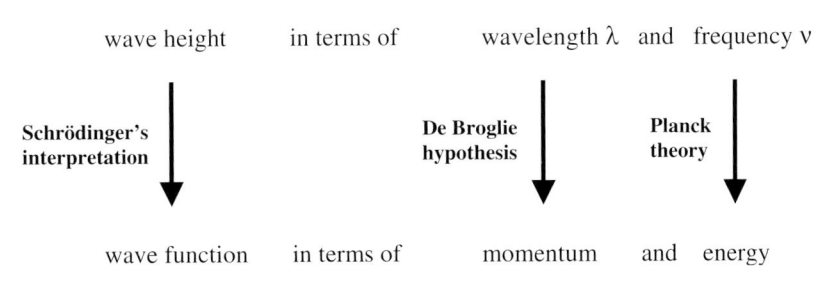

	wave height	in terms of	wavelength λ and frequency ν
Schrödinger's interpretation ↓		De Broglie hypothesis ↓	Planck theory ↓
	wave function	in terms of	momentum and energy

Figure 12.3 The origin of the Schrödinger wave equation.

oscillators that could only possess and radiate energy in units of $h\nu$. If you imagine a mass on the end of a very long cord forming a pendulum then for small swings the mass will move to-and-fro approximately in a straight line. The form of motion of the mass is then a linear harmonic oscillation. The characteristic of such a system is that the frequency of the pendulum, ν (number of swings per second), is independent of the size of the swing, as long as it is not too large, but the energy of the system increases with the size of the swing. For a classical system there are no restrictions on energy, which can vary from zero upwards as the size of the swing increases. Subjecting a linear harmonic oscillator to analysis by the SWE gives a very different outcome. Now the equation only has solutions for ψ when the energy is $\frac{1}{2}h\nu$, $\frac{3}{2}h\nu$, $\frac{5}{2}h\nu$ and so on, showing increments of energy $h\nu$ corresponding to Planck's quantum of energy for frequency ν. If one of these oscillators jumps from a higher allowed energy state to a lower allowed energy state then the energy it will give up, that might appear as a photon, is $nh\nu$ where n is an integer. The lowest possible energy of this type of system, $\frac{1}{2}h\nu$ called *zero-point energy*, plays an important role in low-temperature physics. According to classical ideas, at the absolute zero of temperature, the lowest temperature that can theoretically exist, material should have no energy of motion. The wave equation shows that even at absolute zero there is some residual energy of motion — the zero-point energy.

Having said that solutions for the *wave function* ψ can be found only for particular energies we now have to consider what the wave function represents. In the mathematical solutions of the SWE the wave functions are sometimes *complex*, meaning that they are expressed in terms of a special kind of number, not met with in everyday life but useful to mathematicians and physicists. We shall ignore this complication and assume that ψ is represented by an ordinary number, positive or negative. The interpretation that is put on ψ is that it is a *probability wave*, something that we will now explain. Let us assume that the particle is an electron, perhaps one in an atom, and we have found ψ as a function of its position. If we now do an experiment to locate the electron as a *particle* by placing a detector in a particular position then the two possible outcomes are either that we detect it or that we do not. The probability at any instant that we find the particle in a tiny volume v surrounding a point, where the wave function is ψ at that point, is $\psi^2 v$. As an alternative interpretation we can say that the fraction of time that the electron spends in the volume v is $\psi^2 v$. The quantity ψ^2 is termed the *probability density*.

Let us see what this means for the harmonic oscillator. According to classical theory a body undergoing a linear harmonic oscillation moves backwards and forwards between two limits, which can be set arbitrarily. In the middle of each swing, halfway between the limits, it is moving fastest and at the ends of the swing, at the limits, it is momentarily stationary as it reverses its direction of motion. The consequence of this is that the probability of locating it is highest at the end of the swings, where it is moving most slowly and spends more time, and is lowest at the middle of the swing, the region it moves through most quickly. The relative probability of finding the particle at various points is shown as the red line in Figure 12.4. Outside the limits of the swing, set at −2 and +2 in this example, the probability of finding the particle is zero.

Now we consider the behaviour for a simple harmonic oscillator according to the TISWE; the example we take is the solution for the energy $\frac{5}{2}h\nu$. Solving the TISWE for this energy gives values of ψ shown as the green line in Figure 12.4 with a corresponding blue probability curve ψ^2; notice that the probability is always positive since minus times minus gives plus. There are notable differences between the red curve corresponding to *classical mechanics* and the blue curve corresponding to *wave mechanics*. The classical curve gives zero probability at the centre, maximum probability at the extremities of the swing and zero outside the limits of the swing. These results are entirely consistent with intuition. The wave-equation result shows a peak at the centre, zero probabilities on either side of the centre, larger peaks near the classical limit of the swing and finite probability outside the classical limits. The results from the SWE are completely non-intuitive.

The world of wave mechanics, sometimes called *quantum mechanics*, is a strange one in which intuition leads one astray and where the results often challenge one's imagination. It should be pointed out

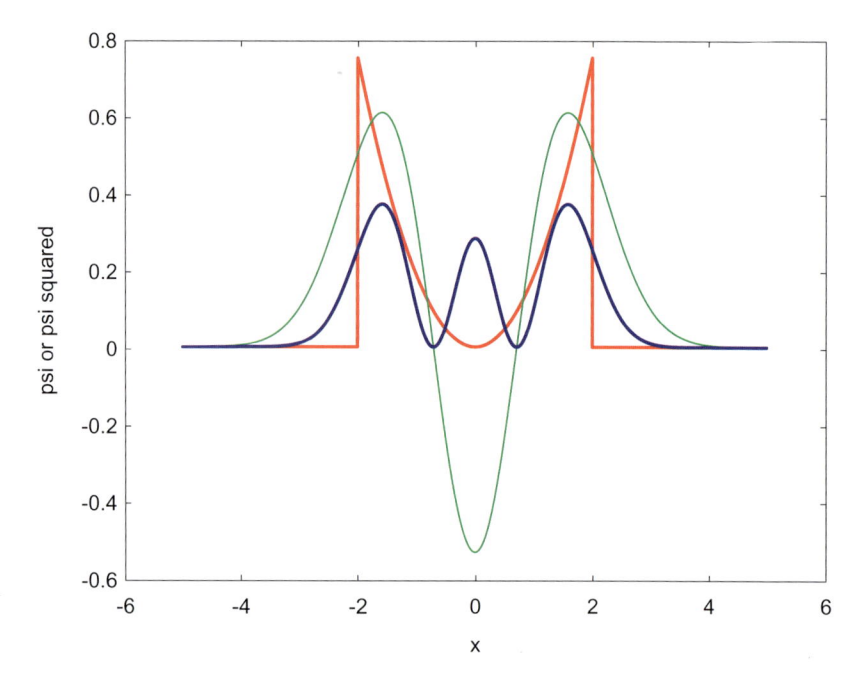

Figure 12.4 A simple harmonic oscillator showing a classical probability curve (red line), the wave function ψ (green line) and the wave mechanics probability curve, ψ² (blue line).

that the validity of Schrödinger's wave equation has not been proved in a strict sense. Although it gives results that may sometimes be difficult to understand, the fact remains that whenever its conclusions are tested against experiment they are found to be correct. As time has gone by, and it has been found to give the right answer over and over again under a wide range of conditions, so confidence in wave mechanics has grown and it is one of the pillars on which modern physics now rests.

12.3 Orbits Become Orbitals

The restriction that Bohr placed on the orbits of electrons gave the right wavelengths for the spectrum of hydrogen but the actual orbits were described by the same classical mechanics that is used to explain the orbits of planets around the Sun. In §11.3 it was shown that the Bohr restriction on the electron orbits could also be put in terms of fitting an exact number of de Broglie waves around the circumference. If the de Broglie hypothesis is accepted, then this makes the restriction on the orbits seem somewhat more logical and certainly aesthetically more satisfying. The hydrogen atom is a simple system with a single particle, an electron, experiencing the electrostatic force due to another particle, the central proton. In a classical treatment of the motion of an electron around a proton, or a planet around the Sun, both bodies actually move. However, since the central body in both cases is so much the more massive then it is possible as a good approximation just to consider the motion of the lighter body and treat the central body just as the unmoving source of a force — gravitational for the Sun and electrostatic for the proton. For this reason it is possible to apply wave mechanics to study the properties of an electron in a hydrogen atom without at the same time applying it to the proton.

When wave mechanics is applied to the electron of a hydrogen atom it is found that solutions for ψ can only be found for particular energies of the electron. Just as for the Bohr model, electrons can jump from one allowed state to another allowed state but cannot exist anywhere in-between. To jump to a higher state the electron must receive energy from an outside source — say by a collision with another atom. If it jumps to a lower state then it will emit a photon with one of the characteristic frequencies of the hydrogen spectrum. So far this looks very much like a description of the Bohr atom in terms of what is observed but, when we consider the wave functions found for the electron, there are important differences in the two descriptions. One obvious difference is that a Bohr (or Sommerfeld) orbit is two-dimensional whereas the electron wave functions are three-dimensional.

The solutions of the wave equation follow the general pattern of the Bohr and Sommerfeld models. Wave functions, that is states of the electron, can be found for solutions associated with three integers n, l and m called *quantum numbers*. The restrictions on the values of l and m are just those shown in Table 11.1 — that is l can vary between 0 and $n - 1$ and m can take the $2l + 1$ integral value between $-l$ and $+l$. The solutions corresponding to the Bohr circular orbits (those with $l = 0$) are radially symmetric, meaning that along any line moving outwards from the nucleus in any direction the wave function changes in the same way. This variation, given as the wave function, is shown as the faint lines in Figure 12.5 for the *principal quantum number* $n = 1, 2$ and 3. The figure also shows as a thicker line the *radial probability density*, which is the probability of finding the electron at

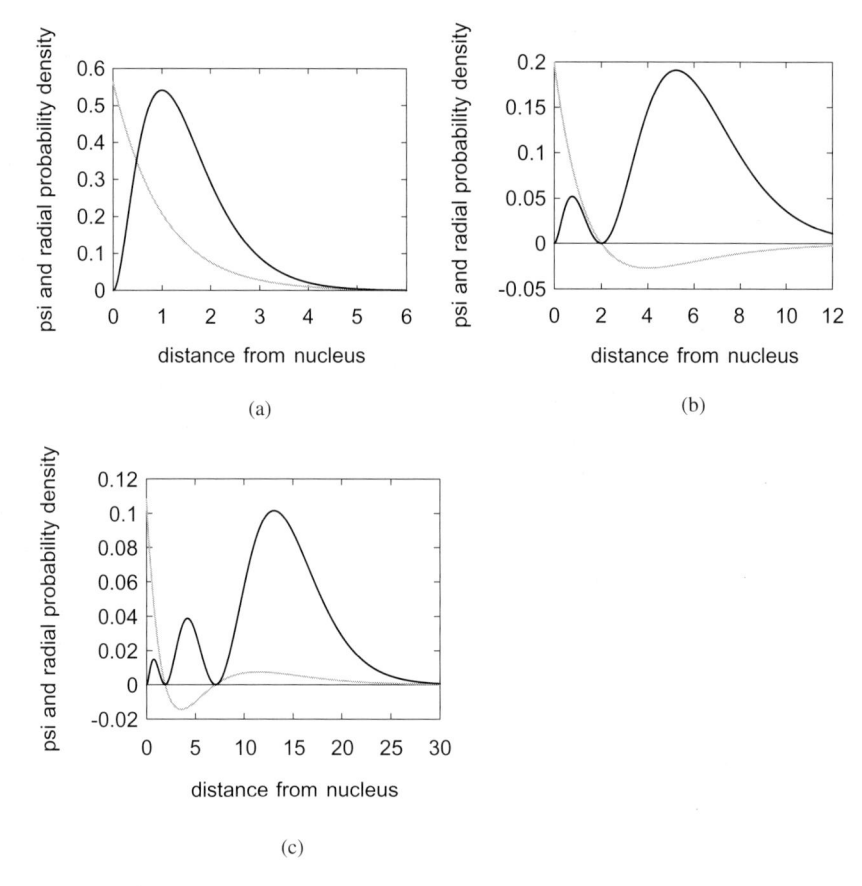

Figure 12.5 The hydrogen radially symmetric wave functions (faint line) and radial probability densities (heavy line) for (a) $n = 1$, (b) $n = 2$ and (c) $n = 3$.

different distances from the nucleus. It depends not only on the probability function ψ^2 but also on the distance from the nucleus, since at greater distances from the nucleus there is a greater volume contained in any small fixed distance interval.

The contrast between these results and the Bohr model is clear. For example, with the Bohr model in the ground state there is only one distance from the nucleus at which the electron can be found. For the wave-equation model it can be found over a wide range of distances, although the most probable distance corresponds to the Bohr-model distance. For higher values of n there are n peaks in the radial probability density although the highest peak moves outwards with increasing n. Again the results conflict with intuition; for $n = 2$ the distance from the nucleus for the Bohr theory is four of the distance units shown in Figure 12.5. The wave-mechanics solution gives the largest radial probability density further out than that but also shows a smaller probability peak further in — not something that would be anticipated.

The wave functions of the electrons of hydrogen, or indeed for the electrons of any kind of atom, are referred to as *orbitals*, a term which sounds similar to, but must not be confused with, *orbits*.

A feature of wave mechanics is that *every* property of a particle is defined by its wave function; we have already seen that the wave function is linked to the energy of the particle. Another property that a particle can possess is angular momentum, as defined in §11.1 and in terms of which the Bohr orbits were originally defined. In the Bohr model an electron moves round in a circular orbit and this moving negative charge constitutes a current. A current moving in a closed loop generates a magnetic field and so, as previously mentioned in §11.4, an electron in an orbit behaves like a small magnet. Although in the wave function the conceptual picture of an electron creating a current in a ring is lost, nevertheless wave functions do define angular momentum, with associated magnetic properties. However, here is an interesting point — from the solutions of the wave equation, the radially symmetric orbitals corresponding to the Bohr circular orbits ($l = 0$) have *zero* angular momentum and yet it was through the finite values of their angular momentum that these Bohr orbits were defined!

In looking for solutions of the wave equation for hydrogen, if the condition is imposed that the solution has to be radially symmetric ($l = m = 0$) then just fixing one integer n, which defines the energy of the electron, gives orbitals, the first three of which are shown in Figure 12.5. The more general solutions, involving the *azimuthal quantum number l* and the *magnetic quantum number m*, are not radially symmetric and the way they vary with distance from the nucleus depends upon direction as well as distance from the nucleus.

The solutions of the wave equation for the electron of hydrogen can be found in the form of exact equations involving the three quantum numbers. However, for more complicated atoms, where the effect of electrons on each other must be taken into account, the form of the wave equations can only be found by numerical methods, which do not give exact answers but are precise enough to confirm that wave mechanics is giving solutions that agree with observations.

12.4 The Great Escape

A very common form of radioactivity involves the emission of an α particle and the interpretation of this is that there are assemblages of two protons and two neutrons forming small α-particle subsystems within the nucleus. From a classical point of view, due to the repulsive forces between like charges, a nucleus should fly apart and the fact that it does not do so means that there is some counterbalancing force. Protons and neutrons are jointly referred to as nucleons. When nucleons are very close together, as in the nucleus, then an attractive force operates, called the *strong nuclear force*, which is stronger than electrostatic forces and binds the nucleons together. So, given the operation of the strong nuclear force the question then changes to how it is that an α-particle ever manages to leave the nucleus in a radioactive atom. Clearly it is not something that happens easily since an atom, such as uranium, may exist for several billion years before the α-particle escapes. Given that it stays so long within the nucleus it then seems strange that suddenly it is able to break out of captivity.

Figure 12.6 Kicking a ball over a hill.

To consider this problem we first consider a very classical situation illustrated in Figure 12.6. The objective of the footballer shown is to kick the football over the hill. If he doesn't kick the ball hard enough then the ball will go up the hill a short distance and then return to him moving in the opposite direction but with the same speed as when he kicked it. The effect of the hill is to act as a reflector, sending the ball backwards with the same speed as it initially went forward. However, if the ball is kicked sufficiently hard then it will slow down until it reaches the top of the hill and then accelerate down the hill on the opposite side. By the time it reaches the point A, at the same level as the footballer, it will be moving at the same speed as when it started its journey.

We now consider the problem of an α-particle escaping from a nucleus, illustrated in Fig. 12.7. The α-particle is moving within the nucleus as shown by the arrow. The effect of the strong nuclear force is similar to that of placing the particle within a well; once outside the well the electrostatic forces take over. If only the α-particle could get out of the well it would then roll down the electrostatic-force hill and escape from the nucleus. However, since its energy of motion cannot get it to the top of the well then, just like the football, whenever it hits the side of the well it reverses its motion until it is reflected from the other side of the well. It rattles around inside the nuclear well, unable to escape. When an α-particle is observed coming from a radioactive atom its speed is a considerable fraction of the speed of light so, from the analogy of the football getting over the hill, we would expect that it was moving inside the nucleus with similar speed. Even if the half-life of the radioactive atom were a hundredth of a second the α-particle must strike the sides of the well about 10^{20} times

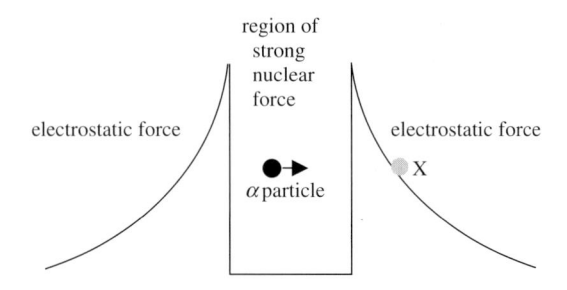

Figure 12.7 An α-particle trapped within a nucleus.

on average before it escapes, so the obvious problem is to understand how it eventually does get out.

The answer to this puzzle comes from wave mechanics and the clue to the solution is found in Figure 12.4. From a classical point of view every time the α-particle strikes the edge of the well it is reflected because classical mechanics forbids it to enter the region outside the well. However, Figure 12.4 shows us that in the world of wave mechanics there is a possibility that the particle can exist outside the classical region. When Schrödinger's equation is solved for the α-particle the wave function, and hence probability density, has a very small value in the region outside the well and hence there is a tiny chance that it can burrow its way through the wall of the well and escape. It is as though the α-particle shown in Figure 12.7 has somehow drilled its way through the force barrier and appeared at the point X. This phenomenon is known as *tunnelling*. Although the probability of a tunnelling event per collision with a wall is extremely small, the total number of collisions per unit time taking place, due to the vast number of atoms in a discernable sample of a radioactive material, is very large so there is a finite rate of radioactive emission from the sample as a whole. Actually, tunnelling is a very important phenomenon in everyday life, since many modern electronic devices depend on tunnelling for their operation.

12.5 Heisenberg and Uncertainty

No account of the new mechanics would be complete without reference to the theoretical physicist, Werner Heisenberg (Figure 12.8),

Figure 12.8 Werner Heisenberg (1901–1976).

previously mentioned in §10.4 in relation to his work that confirmed the existence of the neutron as a distinctive particle. At the same time as Schrödinger was developing his wave mechanics based on the transformed wave equation, Heisenberg was developing an alternative approach based on a branch of mathematics that uses two-dimensional arrays of numbers, called *matrices*. This approach is formally equivalent to that of Schrödinger although, in general, more difficult to apply. Another important theoretical contribution by Heisenberg is very widely known as the *Heisenberg uncertainty principle*. This leads to the conclusion that it is impossible in practice to know exactly both where a particle is and how it is moving. The more precisely one finds the position then the less precisely one knows how it is moving — and *vice versa*. It follows from this principle that it is impossible exactly to define the state of the universe at any time and hence that it is impossible to predict the future state of the universe. This conclusion, of great philosophical significance, contrasts with the determinism of classical physics that would, at least in principle, allow the future evolution of the universe to be deduced from its present state.

Chapter 13

Electrons and Chemistry

13.1 Shells and the Periodic Table

When Mendeleev set up the Periodic Table he had no knowledge of the way that atoms were constructed, with a compact nucleus consisting of protons and neutrons and with external electrons giving the atom both its volume and its chemical properties. Now that we know something about atomic structure, in particular the existence of electrons, we can explain the principles behind the structure of the Periodic Table. There are two basic rules that operate in explaining the way that the electrons are arranged in an atom.

(Rule 1) For any set of quantum numbers, n, l and m, there can be up to, but no more than, two electrons. This rule seems arbitrary but we shall discuss it further and make it seem more rational in §14.3.

(Rule 2) The arrangement of electrons in any atom is such as to give the minimum possible total energy. That is a general principle for any physical system — the state of stable equilibrium, which it will retain if not disturbed, is a state of minimum energy.

The arrangement of electrons is best described by introducing the idea that the electrons exist in shells corresponding to different energies. The first three shells are listed in Table 13.1 together with part of the fourth shell.

Table 13.1 The relationship of shells to the quantum numbers of orbitals. Different shells are separated by the horizontal lines.

Shell	n	l	m	Number of Electrons
1	1	0	0	2
2a	2	0	0	2 ⎫ Combined
2b	2	1	−1	⎫ ⎬ shell 2
	2	1	0	⎬ 6 ⎭
	2	1	1	⎭
3a	3	0	0	2 ⎫ Combined
	3	1	−1	⎫ ⎬ shell 3ab
3b	3	1	0	⎬ 6 ⎭
	3	1	1	⎭
3c	3	2	−2	⎫
	3	2	−1	⎪
	3	2	0	⎬ 10
	3	2	1	⎪
	3	2	2	⎭
4a	4	0	0	2

The first thing that we notice is that, in compliance with rule 1 given above, the number of electrons in a shell is twice the number of different sets of quantum numbers in that shell. The first shell, with $n = 1$, contains two electrons. The second shell is in two sub-shells, 2a and 2b, but for our present purpose we can regard it as a single shell with 8 electrons. The shell with $n = 3$ has three sub-shells, 3a, 3b and 3c, containing 2, 6 and 10 electrons respectively but again we combine the two sub-shells 3a and 3b to give 3ab containing 8 electrons. In the table we also show sub-shell 4a, the first with $n = 4$, which contains 2 electrons. There are also three other sub-shells with $n = 4$, not shown, containing 6, 10 and 14 electrons respectively.

We now illustrate how the electronic structure leads to the Periodic Table, a process that can be followed in Figure 13.1. As indicated by rule 2, the general principle that operates in building up the

Electrons and Chemistry 123

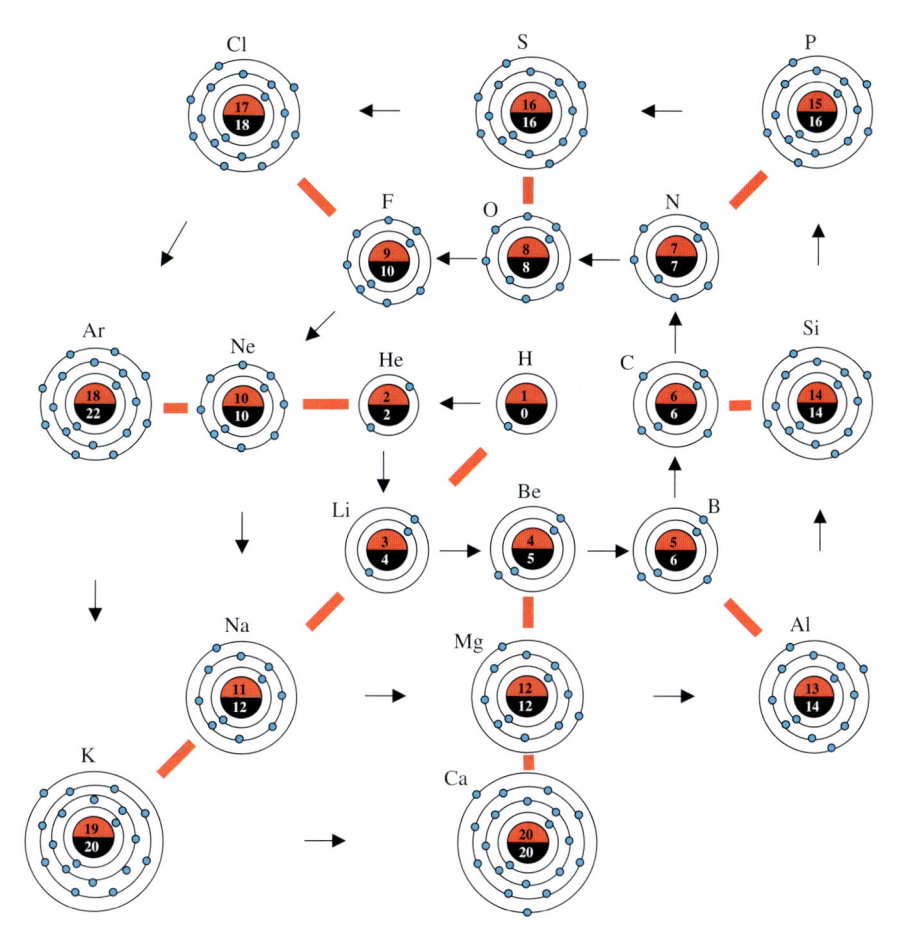

Figure 13.1 A representation of the Periodic Table from hydrogen to calcium. The numbers in the red and black semicircles are the numbers of protons and neutrons respectively. The electrons (blue circles) are shown in the shells 1, 2, 3ab and 4ab. The progression of atomic numbers is shown by the black arrows and elements with similar chemical properties are linked by red bars. The chemical symbols are: hydrogen (H), helium (He), lithium (Li), beryllium (Be), boron (B), carbon (C), nitrogen (N), oxygen (O), fluorine (F), neon (Ne), sodium (Na), magnesium (Mg), aluminium (Al), silicon (Si), phosphorus (P), sulphur (S), chlorine (Cl), argon (Ar), potassium (K) and calcium (Ca).

electronic structure of an atom is that, given the number of electrons equals its atomic number, it has a configuration of lowest energy — the state of greatest stability.

We start with the simplest atom hydrogen, consisting of one proton and one electron in shell 1. The next element is helium with a nucleus consisting of two protons and two neutrons (equivalent to an α particle) and with two electrons that fill shell 1. For lithium, with atomic number 3, the third electron occupies shell 2 and then, as we move through beryllium, boron, carbon, nitrogen, oxygen, fluorine and neon, with each increase of atomic number by one there is one more proton in the nucleus (with some extra neutrons) and an additional electron in shell 2 until, when we reach neon with atomic number 10, shell 2 is full. Now with sodium we begin to fill shell 3ab until, when we reach argon with atomic number 18, shell 3ab is full. The red bars link elements with similar chemical characteristics and it will be seen that the common feature of these sets of elements is that they have the same number of electrons in their outermost shells. Thus hydrogen, lithium, sodium and potassium all have a single electron in their outermost shells.

The general pattern of filling up successive shells is disturbed when we get to potassium. The reason for this is that the sub-shell 3c electrons, with $n = 3$ and $l = 2$, have rather extended orbitals; this can also be seen in the corresponding Sommerfeld orbit shown in Figure 11.7. These extended orbitals have greater energy than the spherically symmetric and rather more compact orbital corresponding to sub-shell 4a and hence the next electron added after argon to give the element potassium (symbol K) is in shell 4a. For calcium, the next element after potassium, the added electron also goes into 4a which then fills that sub-shell. If we progress to the next element, scandium (not shown in Figure 13.1), the added electron starts filling up sub-shell 3c that, for potassium and calcium was left empty, since the energy for this sub-shell is less than that for shell 4b, and as we progress from scandium we progressively fill the ten electron slots in sub-shell 3c.

In presenting this picture of building up the periodic table we have referred to sub-shells with the letters a, b, c and so on. For historical reasons, scientists use the letters s, p, d, f and so on instead.

All sub-shells of type s have two electrons, sub-shells of type p have six electrons and sub-shells of type d have ten electrons. Thus, using this notation, the electronic structure of sodium is $1s^2 2s^2 2p^6 3s^1$ that we translate as follows:

$1s^2$ means that the only sub-shell of shell 1, 1s, has its full complement of 2 electrons
$2s^2$ means that the lowest sub-shell of shell 2, 2s, has its full complement of 2 electrons
$2p^6$ means that the other sub-shell of shell 2, 2p, has its full complement of 6 electrons
$3s^1$ means that there is 1 electron in the lowest sub-shell of shell 3, 3s.

In the same notation the electronic structure of calcium is $1s^2 2s^2 2p^6 3s^2 3p^6 4s^2$ indicating that the sub-shell 3d is empty but that the sub-shell 4s has its full complement of two electrons.

13.2 Valency

The chemical significance of the shell structure is that it is the extent to which the outermost shell is occupied that dictates the chemical behaviour of an atom. In Figure 13.1 we notice that the three related elements, helium, neon and argon, have the common characteristic that their outer shells, or major sub-shells, are full. These elements are the *noble gasses* that share the characteristic that they are chemically inert — that is to say that they do not take part in reactions with other elements. The reason for this is that with their outer electron shells full they are in a state of lowest energy that cannot be made lower by combination with another element. We now consider a different situation with the two elements sodium and chlorine, the electronic structures of which are shown in Figure 13.2. Sodium has a lone electron in its outer shell and if it could lose this it would be left with a complete outermost shell. By contrast, chlorine has seven electrons in its outer shell and would require one more to make its shell complete. This is the situation for a perfect marriage. If they get together and sodium donates its electron to chlorine then they will both share the

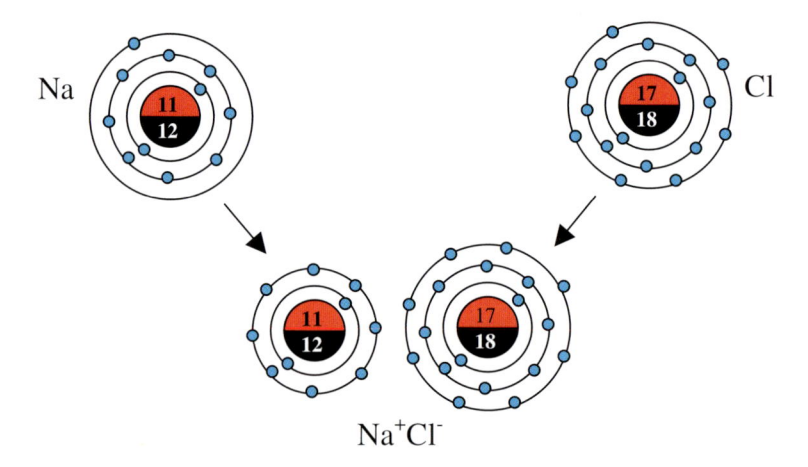

Figure 13.2 The formation of sodium chloride by ionic bonding.

benefit of having a completed outer shell, the state that every atom would like to achieve. The total energy of the electrons in the sodium and chlorine atoms is less after the transfer of the electron was made than it was before.

In giving up its electron to chlorine the sodium atom acquires a positive charge while at the same time the chlorine atom, with an extra electron, has become negatively charged. The opposite charges give a force of attraction that binds the sodium and chlorine together. Actually, as is illustrated in Figure 18.14, sodium chloride forms a framework structure but the forces binding neighbouring atoms together in the framework are due to the transfer of electrons from sodium to chlorine atoms. Atoms that lose or gain electrons to acquire a charge are referred to as *ions*, so the type of bonding that produces sodium chloride is called *ionic bonding*.

The number of electrons that an atom has either to lose or to gain to give a complete shell or major sub-shell defines its *valency*, so sodium has a valency of 1 and chlorine also has a valency of 1. The two cases can be distinguished by describing sodium as *electropositive*, meaning that it has an electron to contribute, and chlorine as *electronegative*, meaning that it is a willing recipient of an electron. There can also be ionic bonding where, for example, an atom of

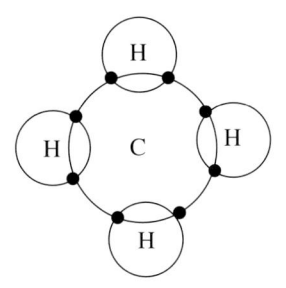

Figure 13.3 The molecule of methane held together by covalent bonds.

valency 2 combines with two atoms of valency 1. In the case of magnesium chloride the magnesium of valency 2 gives one electron to each of two atoms of chlorine to form the $MgCl_2$, another framework structure held together by ionic bonding.

There is another major kind of bonding, *covalent bonding*, which depends not on the donation of electrons but on the *sharing* of electrons. As an example in Figure 13.3 we show the formation of methane, CH_4, by the sharing of electrons. Carbon has a valency of 4, having four electrons in its outer shell. By sharing its electrons with each of the four hydrogen atoms, as shown, both the carbon atom and the hydrogen atoms derive the benefit of having closed outer shells — the hydrogen atoms acquire an extra electron in their 1s shells while carbon gains four electrons to fill its 2p shell.

Other kinds of bonding do occur but what we have described here are the two major types. They illustrate the basic principles that govern the way that chemical compounds form and the role that electronic structure plays in their formation.

Chapter 14

Electron Spin and the Exclusion Principle

14.1 The Stern–Gerlach Experiment

In 1922 the German (later American) physicist Otto Stern (1888–1969; Nobel Prize for Physics, 1943) and the German physicist Walther Gerlach (1889–1979) designed and carried out a famous experiment. The experimental arrangement is illustrated in Figure 14.1. Heating silver in a high temperature furnace, with a small hole through which the vaporized silver could leave, produced a beam of silver atoms. These were then passed through two slits that gave a narrow band of silver atoms streaming into an evacuated space. The silver atoms then passed through a magnetic field between the oddly shaped pole pieces of a magnet. Finally the silver atoms impinged on a screen where their positions could be recorded.

The idea behind the oddly shaped pole pieces is that they gave a field that was non-uniform. The effect of this on a magnet placed in the field is that the magnet experiences a force either in or opposed to the direction of the field; the theory behind this effect is illustrated in Figure 14.2.

The equivalent magnet associated with an atomic orbital behaves just like a real bar magnet, which consists of separated north and south poles. In the figure the north pole (black) experiences an upward force and the south pole (grey) a downward force. Because the field is non-uniform the magnitudes of the forces are different and, in the case shown in the figure, it is the upward force that is

130   *Materials, Matter and Particles*

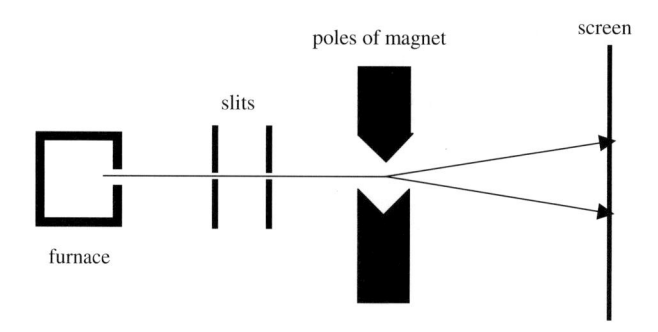

Figure 14.1 The Stern–Gerlach experiment. The beam of silver atoms splits into two beams.

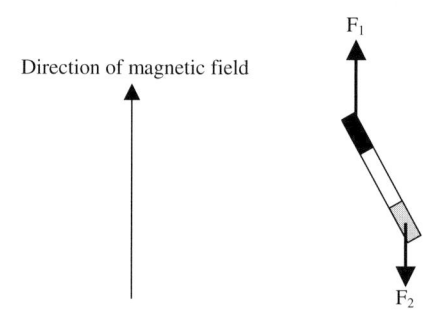

Figure 14.2 In a non-uniform field the forces on the two poles of the magnet are unequal.

the stronger. This means that the magnet will move in an upward direction.

When the result of the Stern–Gerlach experiment — the splitting into two beams — was first observed it was difficult to understand. Silver has atomic number 47 and a valency of 1 because the electron structure consists of a number of full shells plus the single valence electron, the orbital of which is 5s. In the closed shells, for every electron with quantum numbers n, l and m there is another with quantum numbers n, l–m. The forces due to the external field on the equivalent magnets of this pair of orbitals cancel each other along the magnetic field direction so they would contribute no net force on the silver atom. Since the closed shells give a zero force and a 5s electron has zero angular momentum (true for any orbital with $l = 0$), and hence no associated equivalent magnet, it was thought that a silver atom should be completely non-magnetic and hence be entirely unaffected by a magnetic field.

A new theory was needed to explain these results and it was not long in coming.

14.2 Electrons in a Spin

In 1925 two Dutch, both later American, physicists, Samuel Goudsmit (1902–1978) and George Uhlenbeck (1900–1988) proposed that the explanation of the Stern–Gerlach experiment was that an electron has an intrinsic spin, with an associated angular momentum and magnetic moment. There was a problem with this suggestion. From the allowed values of the magnetic quantum number m the number of possible orientations of the equivalent magnet is $2l + 1$ and a stream of particles, each with azimuthal quantum number l, passing through a non-homogeneous magnetic field, would be split into $2l + 1$ separate beams. The difference of the component of angular momentum in the direction of the magnetic field from one magnetic quantum number to the next was $\hbar$. Applying this to the two beams of the Stern–Gerlach experiment implies that $2l + 1 = 2$ or $l = \frac{1}{2}$, seemingly inconsistent with quantum ideas. Could it be that electrons have a spin that is one-half of the basic unit of spin, $\hbar$, a basic unit which came from Bohr's theory and as had naturally arisen from the application of wave mechanics? A very prominent Austrian physicist, Wolfgang Pauli (Figure 14.3; Nobel Prize for Physics, 1945) was initially very opposed to the whole idea of half-integral spin but later changed his mind. It was mentioned in §12.4 that Heisenberg had developed an alternative approach to quantum mechanics based on the use of matrices. While this approach was, in general, more difficult to apply than Schrödinger's wave equation, when it came to describing angular momentum it was ideal. Pauli was able to describe the phenomenon of electron spin in terms of what are called *Pauli spin matrices* from which the result that electrons should have half-integral spin comes about in a natural way.

The idea of electron spin now added a new quantum number. The state of each electron in an atom can be described in terms of four

Figure 14.3 Wolfgang Pauli 1900–1958).

quantum numbers — the principal quantum number, n, the azimuthal quantum number, l (values from 0 to $n-1$), the magnetic quantum number, m (values from $-l$ to $+l$) and a *spin quantum number*, s, which could take one of the two values $+\frac{1}{2}$ and $-\frac{1}{2}$. The valence electron of the silver atoms in the Stern–Gerlach experiment, passing through the magnetic field had quantum numbers either $(5, 0, 0, +\frac{1}{2})$ or $(5, 0, 0, -\frac{1}{2})$ and the two kinds were deflected in opposite directions to give the two beams.

It is sometimes asked why it was that silver atoms were used for the Stern–Gerlach experiment — why not isolated electrons since beams of electrons can be produced? The answer to this question is given by Heisenberg's uncertainty principle (§12.4), the consequences of which become greater as the mass of the particle being observed becomes smaller. Those engaged in sports do not have to worry about the uncertainty principle since the uncertainty in the position and speed of a ball is completely negligible compared to its actual position and speed. In passing electrons through a magnetic field the uncertainty would operate in such a way that the separated beams, as seen in Figure 14.1, would be so smeared out that they would appear to be a single blurred beam. Increasing the mass associated with the electron spin, by using silver atoms, requires the use of a more non-uniform and stronger magnetic field to give a measurable splitting of the beams but this can be achieved while, at the same time, the uncertainty principle does not prevent the splitting from being observed.

14.3 The Pauli Exclusion Principle

The concept of electron spin gave the final component of a complete and logical explanation of the electronic structure of atoms and the structure of the Periodic Table. In 1925 Pauli proposed an idea, now known as the *Pauli exclusion principle*, one conclusion from which is that all the electrons in an atom must have a different set of four quantum numbers. We can imagine our atom situated in a magnetic field, so that the magnetic quantum numbers associated with angular momentum have some meaning. The exclusion principle then places a strong restriction on the electronic structure. As an example we consider an oxygen atom in its ground state so that electrons occupy the lowest energy states they can, consistent with the exclusion principle. The lowest energy state is with $n = 1$. The only allowed value of l is zero (l must be less than n) and hence $m = 0$ (m can have all values from $-l$ to $+l$). However, the values of n, l and m place no restriction on the value of s which can have values $+\frac{1}{2}$ and $-\frac{1}{2}$. Following the logic of this approach the electronic structure of oxygen is

n	l	m	s
1	0	0	$+\frac{1}{2}$
1	0	0	$-\frac{1}{2}$
2	0	0	$+\frac{1}{2}$
2	0	0	$-\frac{1}{2}$
2	1	0	$+\frac{1}{2}$
2	1	0	$-\frac{1}{2}$
2	1	1	$+\frac{1}{2}$
2	1	1	$-\frac{1}{2}$

We now refer back to the description of the shell structure of atoms given in §13.1. For $n = 1$ there are just two electrons in the shell, allowed by the exclusion principle because there are two electron-spin states. In the shell-structure description of oxygen this appears as $1s^2$, the '1' giving the value of n, the 's' (not to be confused with s signifying electron spin) indicating that $l = 0$ and the '2' giving the number of electrons in that shell, in this case filling the shell. For $n = 2$ there are two electrons with $l = 0$ so that the sub-shell is completely full and the shell description is $2s^2$. However, there is another sub-shell, that with $l = 1$, and this could contain up to six electrons, the four given in the list above plus two others with quantum numbers $(2, 1, -1, +\frac{1}{2})$ and $(2, 1, -1, -\frac{1}{2})$. The content of this sub-shell is described as $2p^4$, where the 'p' indicates $l = 1$. This gives the description of oxygen as $1s^2 2s^2 2p^4$. Oxygen has a valence of 2 since it is a willing recipient of two electrons to complete the shell with $n = 2$. Thus it can combine with magnesium, also of valency 2, to give magnesium oxide, MgO, or with two atoms of sodium, of valency 1, to give sodium oxide, Na_2O.

Although the Pauli exclusion principle has been introduced as though it were a fairly arbitrary rule that happened to explain observations, it does actually have a very sound theoretical foundation. Electrons are a particular kind of particle that has half-integral spin but there are others, protons and neutrons, for example, that also have half-integral spin. Such particles are referred to collectively as *fermions*, named after an Italian physicist Enrico Fermi, about whom we shall say more later (§17.6). There are other particles that are met with, α particles for example, that have integral spin, meaning that their spin angular momenta can either be zero or be some integral number of $\hbar$. Such particles are called *bosons*, named after an Indian theoretical physicist Satyendra Nath Bose (1894–1974). The two kinds of particles have different properties, in the sense that if there are collections of large numbers of the particles at a given temperature, then the distribution of energies of the individual particles is different in the two cases. The distribution for fermions is governed by *Fermi-Dirac statistics* while that for bosons is governed by *Bose-Einstein statistics*. We need not concern ourselves with the details of the difference, except to note that there is one. The intricacies of particle physics are many — and most of them are best left to the specialists.

Chapter 15

Isotopes

15.1 What is an Isotope?

Figure 13.1 shows a representation of part of the Periodic Table. For the nucleus of each element there is given the number of protons, that governs the kind of atom it is, and the number of neutrons, that add mass but no charge to the nucleus and are necessary to provide the short-range nuclear force that prevents the nucleus from flying apart. Also given are the electrons that, taking into account their shell structure, govern the way in which the atom is involved in chemical reactions. Thus, in the figure we see that a carbon atom has six protons, six neutrons and also six electrons that are arranged in two shells. However, Figure 13.1 is incomplete. Carbon occurs in all living material, and some non-living material, and if we were to examine carefully any of this carbon we should find that just over one per cent of the carbon atoms differ from the description given above. These carbon atoms contain seven, rather than six, neutrons in the nucleus. So, are these still carbon atoms or are they something else? The answer is that they *are* carbon atoms, but a different *isotope* of carbon from that shown in Figure 13.1.

The feature that exclusively identifies the type of atom is the number of protons it contains in its nucleus. The number of neutrons can vary, although usually not by much, and each different number of neutrons gives a different isotope. Again, an atom can lose one or more of its electrons and this will still not change the fundamental description of the atom. A carbon atom at a high temperature or being bombarded by other particles may lose one of its electrons and become a *singly ionized* carbon atom. An even higher temperature, or more severe bombardment, may remove two electrons and cause it to be doubly ionized and more and more electrons can be removed until, when all the electrons are removed one would be left with a carbon nucleus. But, at all stages of this process we basically have carbon because there are the six protons in the nucleus.

Clearly there needs to be some way of describing different isotopes and in the case we have here we can refer to the carbon isotopes as carbon-12 and carbon-13, where the number gives the combined number of protons and neutrons. Protons and neutrons, collectively referred to as *nucleons*, have approximately the same mass so the number is also an indication of the mass of the nucleus. Another notation for describing these two isotopes is $^{12}_{6}C$ and $^{13}_{6}C$ where the top digit on the left-hand side of the C is the mass and the bottom digit the number of protons. From what we have previously said there is some redundancy in this description since C means carbon and for carbon there *must* be six protons. For that reason sometimes the isotopes are indicated in the form ^{12}C and ^{13}C with the six removed — but we shall always use the redundant form.

The two isotopes of carbon, $^{12}_{6}C$ and $^{13}_{6}C$, are stable, that is to say that they are not radioactive and hence do not spontaneously transform into something else. There are other isotopes of carbon that *are* radioactive and one of these will be discussed in the following chapter.

15.2 The Stable Isotopes of Some Common Materials

The simplest atom, hydrogen, exists in another stable form where the nucleus contains a single neutron accompanying the proton. This isotope of hydrogen is so important in many fields of science that it is given the distinctive name *deuterium* and is represented by its own letter symbol D. We shall see in Chapter 24 that deuterium is a significant product of the process whereby matter was first formed in the

Figure 15.1 A hydrogen and deuterium atom showing protons (grey), neutron (black) and electrons (white).

Universe. It is fairly common — for every 100 000 hydrogen atoms on Earth there are 16 atoms of deuterium. It is a component of water so that when your bath is filled it contains several grams of deuterium — and so do you for that matter! Figure 15.1 shows representations of hydrogen and deuterium atoms.

An unusual feature of the stable isotopes of hydrogen is their high mass ratio, 2:1. There are some contexts in which this ratio plays an important role. For example, the planet Venus is a very arid planet — it lost most of its water at some stage of its history — but the proportion of deuterium in the hydrogen it contains is 100 times greater than on Earth, so that 1.6% of its hydrogen is deuterium. This rich deuterium content can be linked to the process by which Venus lost its water.

The early Venus contained a great deal of water and, because of its proximity to the Sun, water vapour would have been an important component of its atmosphere. The effect of ultraviolet radiation from the Sun would be to break up water molecules by

$$H_2O \xrightarrow{radiation} H + OH.$$

Two OH (hydroxyl) groups would combine to give H_2O plus O and the oxygen atom would form compounds with other materials. The net effect was to produce free hydrogen in the atmosphere. However, part of the water would be in the form HDO, where a deuterium atom replaced one of the hydrogen atoms, and when this underwent a similar break-up the effect was to release free deuterium in the atmosphere. At any particular temperature the speed with which gas atoms move depends on their mass; the higher is the mass the lower is the speed. Since the mass ratio of deuterium to hydrogen is so high the deuterium atoms in the atmosphere were moving much more slowly than the hydrogen atoms. Hydrogen atoms were moving quickly enough to overcome the gravitational pull of Venus and so escaped from the planet, but the deuterium was retained. In this way, as Venus lost more and more of its water so the ratio of deuterium to hydrogen increased until it reached its present level.

The other component of water, oxygen, also consists of more than one isotope. The normal oxygen contained in water, and that is also in the air we breathe, consists of three isotopes represented as $^{16}_{8}O$, $^{17}_{8}O$ and $^{18}_{8}O$. The standard terrestrial ratios for those isotopes are $^{16}_{8}O:^{17}_{8}O:^{18}_{8}O = 0.953:0.007:0.040$ and this mixture is known as SMOW (Standard Mean Ocean Water). The fact that it is the *terrestrial* ratios that have been mentioned is because the ratios can be very different in meteorites, bodies that originate from elsewhere in the Solar System. This is true for many different elements — that terrestrial and meteorite isotopic ratios can be distinctively different — and the origin of these differences is something that interests meteoriticists, those scientists that study meteorites.[g] Some other atoms showing important difference between terrestrial and meteorite ratios are, with the terrestrial ratios given:

Magnesium	$^{24}_{12}Mg:^{25}_{12}Mg:^{26}_{12}Mg = 0.79:0.10:0.11$
Neon	$^{20}_{10}Ne:^{21}_{10}Ne:^{22}_{10}Ne = 0.905:0.003:0.092$
Carbon	$^{12}_{6}C:^{13}_{6}Ne = 0.989:0.011$.

A representation of the three neon isotopes is shown in Figure 15.2. This element is of interest to meteoriticists since some samples of neon extracted from meteorites are pure neon-22, which is just a 9% component of terrestrial neon.

There is a wide variation in the number of stable isotopes for different elements. The element with the greatest number of stable

---

[g] This topic is described in some detail in the author's book *Formation of the Solar System: Theories Old and New*.

$^{20}_{10}$Ne (90.5%)　　　　　　$^{21}_{10}$Ne (0.3%)　　　　　　$^{22}_{10}$Ne (9.2%)

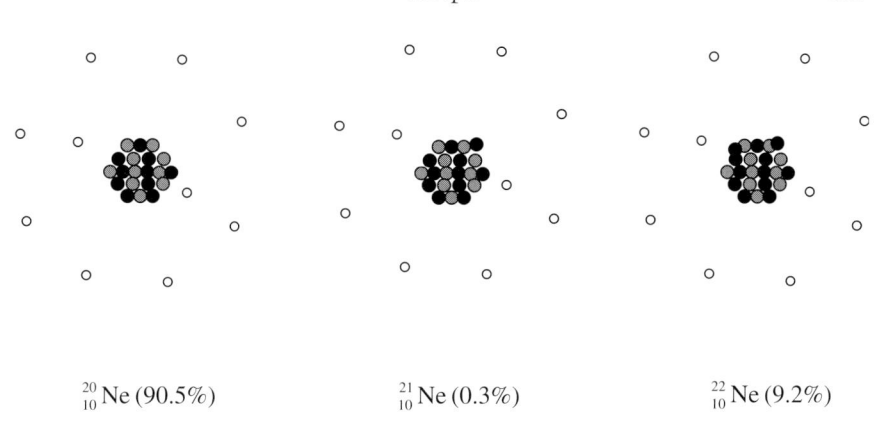

Figure 15.2　The three stable isotopes of neon (particles indicated as in Figure 15.1).

isotopes is tin, atomic number 50, chemical symbol Sn, which has 10 stable isotopes. These are:

$^{112}_{50}$Sn　$^{114}_{50}$Sn　$^{115}_{50}$Sn　$^{116}_{50}$Sn　$^{117}_{50}$Sn　$^{118}_{50}$Sn　$^{119}_{50}$Sn　$^{120}_{50}$Sn　$^{122}_{50}$Sn and $^{124}_{50}$Sn.

There are also elements with only one stable isotope — for example, the only stable isotope of sodium is $^{23}_{11}$Na and of aluminium is $^{27}_{13}$Al, although both those elements have other isotopes that are radioactive, and hence unstable. Finally there are elements for which there are no stable isotopes. For lower atomic numbers, examples are technetium, atomic number 43, chemical symbol Tc, and promethium, atomic number 61, chemical symbol Pm. Actually in a strict sense there are no stable elements beyond bismuth, atomic number 83, chemical symbol Bi. The well-known element uranium, atomic number 92, chemical symbol U, is radioactive but its most common isotope, $^{238}_{92}$U, has a half-life of 4.5 billion years, roughly the age of the Solar System, so about one-half of that existing when the Solar System formed is still around. Many other of the heavier elements fall into the category of having such long half-lives that they still occur in appreciable quantities. By contrast, there are some elements that have such a fleeting existence that they can hardly be considered to exist at all. The element astatine, atomic number 85, chemical symbol At, only occurs as a product from the radioactivity of uranium. The longest-lived isotope of astatine, $^{211}_{85}$At, has a half-life of 8.1 hours so a few days after it has been produced it has virtually disappeared. It is estimated that the total quantity of astatine in the world at any time is about 30 grams — enough to fill a tablespoon!

Our discussion of isotopes has led us to the subject of radioactivity so now we shall consider in more detail the processes that occur to make an atom radioactive.

Chapter 16

Radioactivity and More Particles

16.1 The Emission of α Particles

The discovery of radioactivity by Becquerel and the subsequent work by the Curies was carried out without any understanding of the structure of an atom. The identification of α particles as doubly charged helium atoms by Rutherford, plus the knowledge of the nuclear structure of atoms, suggested that within the nucleus there is some kind of sub-structure in which α particles play a role. In Figure 12.7 we portrayed an α-particle sub-unit trapped within a nucleus, bouncing to-and-fro between its walls until, due to the tunnelling phenomenon allowed by wave mechanics, it manages to break free. The question we now address is what is left behind when it escapes.

The isotopes of uranium found in nature are $^{238}_{92}$U, forming about 99.3% of the total, and $^{235}_{92}$U, comprising most of the remainder, together with slight traces of $^{234}_{92}$U. All uranium isotopes are radioactive. Uranium-238 decays by emitting an α particle to give an isotope of thorium (chemical symbol Th), another element, all of whose isotopes are radioactive. This process is described by

$$^{238}_{92}\text{U} \rightarrow {}^{234}_{90}\text{Th} + \alpha \text{ particle } ({}^{4}_{2}\text{He}).$$

The first thing to notice about this reaction is that it conserves the total number of protons and neutrons. Thorium, with atomic number 90, has two less protons than uranium, with atomic number 92, which is the number of protons in the emitted α particle. Similarly the thorium isotope, with atomic mass 234, has four less nucleons than the uranium isotope, with atomic mass 238, which is the mass of the α particle.

The other two naturally occurring isotopes of uranium are also α-emitters with thorium as the products of the decay. The processes are

$$^{235}_{92}U \rightarrow {}^{231}_{90}Th + \alpha \text{ particle}$$
$$\text{and} \quad {}^{234}_{92}U \rightarrow {}^{230}_{90}Th + \alpha \text{ particle}.$$

Once again it can be seen that there is conservation of the two kinds of nucleons.

The half-life of uranium-238 is 4.47×10^9 years, approximately the age of the Solar System, which means that when the Solar System formed there was twice as much uranium-238 present as there is now. For uranium-235 the half-life is 7.04×10^8 years, indicating that there have been more than 6 half-life periods since the Solar System began. This means that originally there was more than 2^6 (= 64) times as much uranium-235 as there is now. The implication of this is that originally about one-quarter of the uranium on Earth was uranium-235 but, because it has decayed more quickly than uranium-238, it is now less than 1% of the whole.

Another important aspect of a radioactive decay is that heat energy is produced. The combined products of a radioactive decay have a smaller mass than the original nucleus and this loss of mass is transformed into energy in accordance with Einstein's famous equation $E = mc^2$, where E is the energy produced, m is the mass loss and c is the speed of light. This energy is in the form of the kinetic energy of the emitted particles and when it is shared with other material then it is a source of heating. The Earth contains many radioactive materials in its crust, of which uranium is the most important, and to some extent also in the mantle below the crust. The total production of heat by radioactivity, appearing as a heat flow from the surface of the Earth, is at a rate of about 20 terawatts (2×10^{13} watts) — somewhat more than the total world power output of 15 terawatts, produced by fossil fuels (93%), nuclear power (5%) and renewable sources (2%).

Because α particles are so easily absorbed, even by a sheet of paper, if they originate externally they are not a serious health hazard. Even the layers of dead skin that are a normal part of one's anatomy would be enough to prevent them from entering the body to cause damage. However, if ingested an α particle source is extremely damaging to health, and can even be lethal. A landmark in, presumably political, assassination occurred in 2006 when a dissenting former officer of the Russian Federal Security Service, Alexander Litvinenko, living in London, was poisoned with the α-emitter polonium-210, which has a half-life of 138 days. Litvinenko died three weeks after ingesting the polonium.

16.2 β Emission

In stable nuclei neutrons and protons can exist without change in perpetuity. However, an isolated neutron is an unstable particle with a half-life of about 15 minutes. It decays into a positively charged proton and a negatively charged electron, which together have no net charge just like the neutron that produced them, so conserving charge. In some unstable nuclei this breakdown of a neutron can also occur. The effect of this is that a proton replaces the neutron within the nucleus and the electron shoots out of the nucleus at a very high speed and is observed as a β particle. The speeds of these particles can be very high, very close to the speed of light. It should be stressed that these electrons have nothing to do with the atomic electrons arranged in shells around the nucleus — they are from a quite different source.

Replacing a neutron in the nucleus by a proton gives no change in the atomic mass of the atom but the atomic number is increased by 1. Typical β decay is represented by the conversion of caesium-137, with a half-life of just over 30 years, to barium-137, described by

$$^{137}_{55}\text{Cs} \rightarrow ^{137}_{56}\text{Ba} + e^-,$$

in which e^- represents the β particle. Of course, the barium-137 will have to pick up an electron from somewhere to become an electrically neutral atom but there always seem to be spare electrons around — from β-emission, for example.

The emission of very energetic negatively charged β particles is the most common form of β decay but it is not the only kind. However, before discussing the other type of β decay we must first describe a new kind of particle that we have not previously considered.

16.3 The Positron

In 1928 a British physicist, Paul Dirac (Figure 16.1; Nobel Prize for Physics 1933 shared with Erwin Schrödinger) developed a theory of the behaviour of fermions, the spin-$\frac{1}{2}$ particles that were described in §14.3. From the theory it appeared that a particle should exist, similar in all ways to the electron except that it should have a positive charge. This postulated particle was called a *positron*. The electron and positron form a pair of *antiparticles* that, if they come together, would annihilate each other with the lost mass being converted into a pair of γ-ray photons. The need for a pair of photons to be produced rather than a single photon comes from the need to conserve momentum, a necessary conservation in any physical process. The reverse process can also take place. A very high-energy electromagnetic photon interacting with an atomic nucleus can convert some of its energy

Figure 16.1 Paul Dirac (1902–1984).

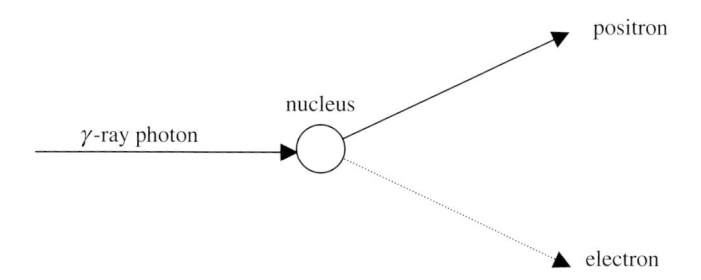

Figure 16.2 Formation of an electron-positron pair by interaction of a γ-ray photon with a nucleus.

into mass in the form of an electron-positron pair. This process is illustrated in Figure 16.2.

The presence of the nucleus is necessary for pair production to take place since there needs to be a body present to take up the extra energy and momentum in the photon that does not go into producing the particles. A consequence of Einstein's equation is that, in order to create a mass $2m_e$, where m_e is mass of the electron (and positron), the energy of the photon must be greater than $2m_ec^2$.

The process of pair production was the agency by which the positron was first identified as a real, rather than just a postulated, particle. In 1932 the American physicist, Carl D Anderson (1905–1991; Nobel Prize for Physics, 1936), placed a lead plate within a piece of apparatus known as a Wilson cloud chamber. This cloud chamber was designed in 1911 by the British physicist, Charles Wilson (1869–1959; Nobel Prize for Physics, 1927) and consists of an enclosure with glass sides containing air saturated with water vapour. When a charged particle passes through the chamber its path becomes visible as a trail of water droplets. Different kinds of particle give tracks with different characteristics. Tracks from α particles are thick and straight and have a short path length indicating how easily they are absorbed. A β particle track is also straight but much finer and longer. If a particle of very low energy enters the cloud chamber then it may show a curved path under the influence of the Earth's magnetic field. A representation of these cloud-chamber tracks is shown in Figure 16.3.

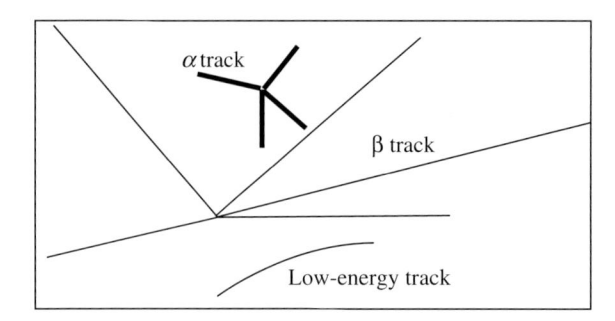

Figure 16.3 Representation of cloud-chamber tracks.

The cloud chamber used by Anderson was constantly being bombarded by *cosmic rays*, as indeed was every other location in the Universe. The term 'rays' in relation to 'cosmic rays' is somewhat of a misnomer since the vast majority of the so-called rays are actually extremely energetic atomic nuclei — mostly protons (hydrogen nuclei) but also heavier nuclei going all the way up to uranium. However, about 0.1% of the total energy of cosmic rays is in the form of very high-energy γ rays and these are what interest us here. A γ-ray photon, with energy greater than $2m_ec^2$ interacting with a lead nucleus in the plate can produce an electron-positron pair and this is what Anderson observed. Actually the cloud chamber tracks from an electron and a positron are identical in appearance but were distinguished by placing the cloud chamber in a magnetic field that causes the tracks from the two particles to bend in opposite directions.

The discovery of the positron was a watershed in understanding the nature of matter. The coming together of an electron and a positron did not give a composite particle with twice the mass of the electron and no charge but rather led to the annihilation of both particles. If the electron had an antiparticle then could a proton and, despite its lack of charge, could a neutron also have antiparticles? Could there be anti-atoms made of the antiparticles of protons, neutrons and electrons and could there be large antimatter bodies somewhere in the universe?

16.4 Another Kind of β Emitter

Since a neutron is intrinsically unstable it is not surprising that in some circumstances it should break up within the nucleus to give β decay. When a neutron disintegrates the products have less mass than the original neutron and this loss of mass provides the energy of the emitted β particle. Clearly, for neutrons present within stable nuclei there is some influence due to the structure of the nucleus that *prevents* it from decaying.

When protons and neutrons come together to form an atomic nucleus then the mass of the nucleus is less than the sum of the masses of the individual nucleons that go to form it. The difference of the masses is called the *binding energy* of the nucleus and it describes the total amount of energy required to break it up into its constituent particles. The masses of nucleons and nuclei are usually given in *atomic mass units* (u) where 1u is defined as one-twelfth of the mass of the nucleus of $^{12}_{6}$C and is $1.660\,539 \times 10^{-27}$ kilogram. The mass of a proton is 1.007 28 u and that of a neutron is 1.008 66 u so the combined mass of two protons and two neutrons, that together form an α particle, is 4.031 88 u. However, an α particle has a mass of 4.001 53 u, less than the combined mass of the components by 0.030 35 u, and the energy equivalent of this difference of mass is its binding energy. If one could break up the α particle into its component parts then this would create extra mass and creating mass requires an input of energy — the binding energy of the α particle.

There are some radioactive isotopes that emit positively charged β particles — energetic positrons. This is done by the conversion of a proton in the nucleus into a neutron and a positron but it is a much more demanding process than the disintegration of a neutron. Since a neutron has more mass than a proton the disintegration of a proton requires an input of energy from some source or other. When a fast positron is emitted by a radioactive nucleus, the required energy to break up the proton comes from a reduction in the binding energy of the nucleus. This reduction in the binding energy of the nucleus must both supply the mass difference between a neutron plus positron and a proton and also the energy with which the positron leaves the nucleus. This kind of radioactive decay is called β⁺ emission as distinct from β⁻ emission when an energetic electron is expelled. By its very nature β⁺ emission is much rarer than β⁻ emission. An example is the decay of radioactive sodium-22 into stable neon-22 by the reaction

$$^{22}_{11}\text{Na} \rightarrow {}^{22}_{10}\text{Ne} + e^+$$

in which e⁺ represents the β⁺ particle. It will be seen that the effect of β⁺ emission is to reduce the atomic number by 1 but leave the atomic mass unchanged.

16.5 The Elusive Particle — The Neutrino

In 1911 the Austrian physicist Lise Meitner (1878–1968) and the German chemist Otto Hahn (1879–1968; Nobel Prize for Chemistry 1945) studied the energies of electrons emitted by β decay by finding how much they were deflected by magnetic fields. Since the same process produces all the β-decay electrons from a particular radioactive atom it might have been thought that they would all have the same energy. Instead, Meitner and Hahn found that the energies varied continuously over a range of values, although there was a distinct maximum energy. This was puzzling since it appeared that energy was being lost in some way and that ran against the important law in physics that in all processes energy must be conserved. There were initially some doubts about the correctness of the Meitner and Hahn results but in the period between 1920 and 1927 they were confirmed by some very precise measurements carried out by a British physicist, Charles Drummond Ellis (1895–1980).

When β decay was observed in a Wilson cloud chamber some even more puzzling results were found. An individual nucleus, emitting a very energetic β particle recoils, just as a gun recoils when it fires a shell. The recoiling nucleus can be seen as a short but thick track in the cloud chamber while the β particle gives its characteristic thin, but long, track. The kind of result found is shown in Figure 16.4.

The pattern of tracks shown in Figure 16.4 was quite unexpected. We have already noted that the varying energy of β particle emission

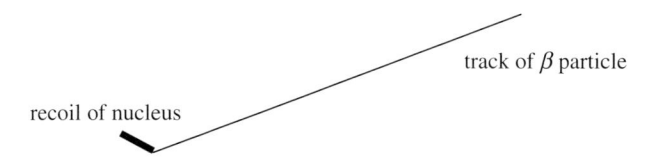

Figure 16.4 Hypothetical tracks of β emission and a recoiling nucleus.

had not been explained but now came the new difficulty that, by all the laws of mechanics, the recoiling nucleus should have moved in a direction exactly opposite to that of the β particle. Actually there was also a third important problem of β decay, tied up with the idea, mentioned at the end of Chapter 14, that protons, neutrons and electrons were all fermions — $\frac{1}{2}$-integral spin particles. Physics has a number of important conservation principles relating to quantities that have to be the same before and after a physical process. One of these relates to angular momentum — which is what spin is all about. The problem was that a fermion, a neutron with $\frac{1}{2}$-integral spin, could decay into two fermions, a proton and an electron, each with $\frac{1}{2}$-integral spin and that did not fit in with the conservation of angular momentum.

Wolfgang Pauli gave the answer to all of these difficulties in 1930. He suggested that there was another particle involved in the break-up of a neutron. These particles have no charge, little mass — or perhaps even no mass, travel at close to the speed of light and are fermions so that they have $\frac{1}{2}$-integral spin. Such particles are able to travel through matter virtually undisturbed and hence they are extremely difficult to detect. The name given to these particles, one suggested by Fermi in 1934, is *neutrino*, which derives from substituting the diminutive ending *ino* on *neutrone*, the Italian word for a neutron. The postulate of such a particle resolved all the problems of β decay. They carried away a variable amount of the energy of β decay so giving a range of energies for the β particles themselves, they would allow the nucleus recoil not to be exactly opposed to the β-particle direction and they solved the spin problem. The three resultant particles from neutron decay — the proton, electron and neutrino — could have a net spin of $\frac{1}{2}$, the original spin of the neutron, because $\frac{1}{2} + \frac{1}{2} - \frac{1}{2} = \frac{1}{2}$.

The neutrino, indicated by the symbol v_e, just like the electron, has an antiparticle, an *antineutrino*, indicated by the symbol $\bar{v}_e$. By convention, the particle produced when a neutron decays into a proton and electron is the antineutrino and the particle produced in conjunction with the break-up of a proton into a neutron and positron is a neutrino. These processes can be described by

$$^1_0 n \rightarrow {}^1_1 p + e^- + \bar{v}_e$$

and

$$^1_1 p \rightarrow {}^1_0 n + e^+ + v_e$$

where $^1_0 n$ and $^1_1 p$ are a neutron and proton, respectively, with their charges and masses indicated in the usual way.

An important characteristic of neutrinos is that, although they are abundant, they are extremely difficult to observe. They can travel through bodies the size of the Earth or the Sun without being absorbed. Vast numbers are constantly being produced in the nuclear processes by which the Sun produces its energy and virtually all of these leave the interior of the Sun and flood out into space. It has been estimated that there are 50 trillion (5×10^{13}) neutrinos passing through every human being per second!

In 1959 two American physicists, Clyde I Cowan (1919–1974) and Frederick Reines (Figure 16.5; 1918–1998; Nobel Prize for Physics, 1995), carried out a very elegant experiment to detect the presence of neutrinos. They used as a source of their neutrinos a nuclear reactor that produced up to one trillion neutrinos per square centimetre per second, a far greater flux of neutrinos than could be produced by any other source. Their experiment depended on an expected reaction where an antineutrino would react with a proton to give a neutron and a positron, a reaction represented by

$$\bar{v}_e + {}^1_1 p \rightarrow {}^1_0 n + e^+.$$

The basis of the experiment was to detect the simultaneous formation of the neutron and positron because this would be a sure sign that the reaction had taken place, and hence that an antineutrino had been present. The experiment is illustrated in Figure 16.6.

Radioactivity and More Particles 151

Figure 16.5 Frederick Reines (1918–1998).

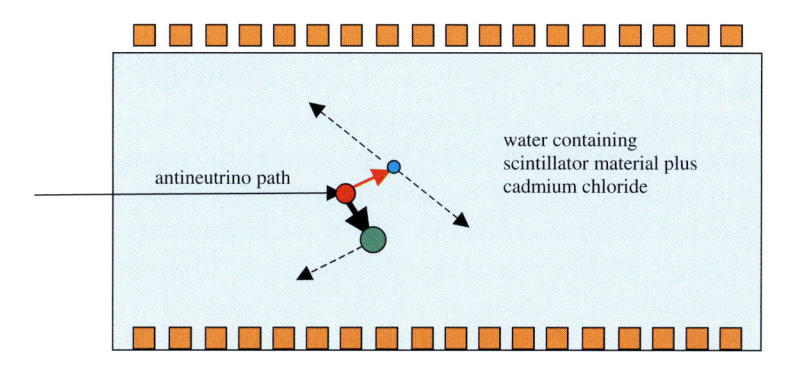

Figure 16.6 A representation of the Cowan–Reines experiment.

The sequence of events that occurred and were recorded by the experiment are as follows:

(a) An antineutrino entered a tank of water within which was dissolved a scintillator — a material that emits flashes of light when γ rays pass through it.
(b) The antineutrino reacted with a proton (red circle) and produced a positron (path represented by red arrowed line) and a neutron (path represented by thick black arrowed line).
(c) The positron interacted with an electron (blue circle) and the resultant annihilation produced two γ ray photons moving in opposite directions (dashed arrowed lines).

(d) The γ rays interacted with the scintillator in the water to give light flashes that were detected by the light detectors (orange squares).
(e) The neutron combined with a cadmium-108 nucleus (large green circle), present in the dissolved cadmium chloride, to give a cadmium-109 nucleus. However, this cadmium-109 nucleus, when first produced, is not in its state of lowest energy so it emits a γ ray and drops to its ground-state energy. This sequence of events is represented by

$$_{0}^{1}n + {}_{48}^{108}Cd \rightarrow {}_{48}^{109}Cd^* \rightarrow {}_{48}^{109}Cd + \gamma$$

in which ${}_{48}^{109}Cd^*$ is the non-ground (excited) state of cadmium-109.
(f) The γ ray (dashed arrowed line) from the cadmium-109 interacted with the scintillator to give a flash of light recorded by the detectors.

The important feature of this experiment is that it relied on the recording of three related events to verify that the neutrino reaction has taken place — first the simultaneous flashes on opposite sides of the apparatus from the positron-electron annihilation event and then, with a delay of five microseconds, the flash from the cadmium process. Individual activation of the light detectors may occur through the action of γ rays from cosmic radiation but the three correlated light detections that indicate a neutrino reaction are unlikely to be anything other than that.

The number of neutrino reactions recorded was about three per hour. As an additional check on the reliability of their results, Cowan and Reines then ran their experiment with the reactor turned off when, as expected, the number of events recorded was significantly less.

16.6 How Old is that Ancient Artefact?

In §15.1 the carbon in living material was described as being mostly carbon-12 with about 1% of carbon-13. Actually a very diligent examination would show another carbon isotope — carbon-14 that is radioactive with a half-life of 5 370 years, present at a level of 1 part in 10^{12} of the total carbon content. Of course, this cannot have been produced when the Earth was formed — it would have long disappeared — and it is present because there is a constant source for its production. High in the atmosphere cosmic rays take part in various nuclear reactions, some of them producing neutrons. These neutrons then interact with the most common isotope of nitrogen to give the reaction

$$^{14}_{7}N + ^{1}_{0}n \rightarrow ^{14}_{6}C + ^{1}_{1}p$$

where $^{1}_{0}n$ is a neutron and $^{1}_{1}p$ is a proton. The radioactive carbon isotope becomes incorporated into the carbon dioxide in the atmosphere and thence into living material. The carbon-14 decays by

$$^{14}_{6}C \rightarrow ^{14}_{7}N + e^{-} + \bar{\nu}_e$$

so restoring the stable isotope of nitrogen from which it originally came. Once the living organism dies, as when a tree is cut down and used to make some artefact, there is no further input of carbon-14 from the atmosphere. The carbon-14 in the artefact decays and hence reduces in amount by a factor of two every 5730 years. Hence the rate at which it emits β^- particles is constantly reducing and from the level of β^- emission per unit mass it is possible to estimate the age of the sample. This is the basis of the carbon-14 dating of ancient organic material; it is able to give reasonably reliable ages up to about 60,000 years, beyond which the rate of β^- emission is too low to give reliable results.

16.7 An Interim Assessment of the Nature of Matter

In looking at ideas about the nature of matter we have seen a fluctuation in the number of different entities by which all matter can be described. Starting with the four basic elements of ancient times — earth, air, fire and water — the advance of chemistry then completely changed the picture by the substitution of a large number of elements, eventually 92 including the four missing unstable ones. However, although the number of elements had been greatly increased, the discovery of the Periodic Table gave a pattern that related groups of elements in a systematic way. Next the physicists stepped in and initially they showed that just three kinds of particle — protons, neutrons and electrons — were the fundamental constituents of all the elements that the chemists had discovered and, in addition, explained the basis of the Periodic Table. This three-particle description of matter was both simple and satisfying.

When physics became involved in the description of matter it brought with it all the baggage of conservation rules that had to be obeyed and mathematical theories that not only explained what was observed but also made predictions about what had not thus far been observed. In looking at the nature of fermions Paul Dirac predicted that positrons should exist and, indeed, so they did. Experimental studies of β decay led to the idea of the neutrino and, sure enough, that was also detected. The fact that both neutron decay and proton disintegration could occur within the nucleus gave the need for the neutrino to have an antiparticle, the antineutrino — yet another particle. Three new particles had been found that were not necessary to describe the structure of atoms *per se* but that, nevertheless, were real particles. So now there were six particles. Is that the total tally? No, it is not — there are many more.

Chapter 17

Making Atoms, Explosions and Power

17.1 The New Alchemy

The first recorded transmutation of one element into another, other than those that happen naturally through radioactivity, was described in §10.3. Although Rutherford was not quite certain what he had done he had, in fact, converted nitrogen into oxygen by the reaction

$${}^{4}_{2}\text{He} + {}^{14}_{7}\text{N} \rightarrow {}^{17}_{8}\text{O} + {}^{1}_{1}\text{p} + \gamma$$

in which ${}^{4}_{2}\text{He}$ is an α particle and ${}^{1}_{1}\text{p}$ is a proton. This was a precursor of many such reactions that were found subsequently, in which one element is changed into another by bombardment, either by a particle or by radiation. These reactions fall into a number of different categories that are now described.

17.2 Reactions with α Particles

Since radioactive materials are readily available α-particle sources, it was not surprising that the first observed nuclear reaction involved bombardment by those particles. There are many reactions that may be induced by α-particle bombardment such as

$$ {}^{4}_{2}\text{He} + {}^{23}_{11}\text{Na} \rightarrow {}^{1}_{1}\text{p} + {}^{26}_{12}\text{Mg}.$$

This can be expressed in more verbal fashion as

$$\alpha \text{ particle} + \text{sodium-23} \rightarrow \text{proton} + \text{magnesium-26}$$

or, in the concise way that nuclear physicists would describe it, as $^{23}\text{Na}(\alpha, p)^{26}\text{Mg}$. In what follows we shall use the first of these three forms, which describes clearly and completely the nature of the reaction and the particles involved.

A characteristic of all nuclear reactions is that they either produce energy or require an input of energy to enable them to happen. A convenient measure of energy in the atomic regime is the *electron volt* (eV), described in Chapter 10. As an example, the energy required to remove the electron from a hydrogen atom in its ground state is 13.6 eV. When we come to events affecting nuclei the energies involved are much higher and are expressed in units of a million electron volts (MeV). In the reaction described above the energy produced is 1.82 MeV. The energy released by a nuclear reaction is referred to as its *Q value* that, if negative, indicates that energy must be provided for the reaction to be able to take place.

Another similar reaction, this time involving the collision of an α-particle with aluminium-27, the only stable isotope of aluminium, is

$$^{4}_{2}\text{He} + ^{27}_{13}\text{Al} \rightarrow ^{1}_{1}\text{p} + ^{30}_{14}\text{Si}$$

in which the final product is a stable isotope of silicon. The Q value for this reaction is 4.01 MeV.

In both of the above reactions the α-particle bombardment gave a proton as one of the final products. Two other kinds of reaction involving α-particle bombardment are illustrated by

$$^{4}_{2}\text{He} + ^{16}_{8}\text{O} \rightarrow ^{20}_{10}\text{Ne} + \gamma.$$

$$^{4}_{2}\text{He} + ^{17}_{8}\text{O} \rightarrow ^{1}_{0}\text{n} + ^{20}_{10}\text{Ne}.$$

In the first of these reactions the α-particle is completely absorbed by the most abundant isotope of oxygen to give the most abundant isotope of neon plus a γ photon. In the second reaction the stable oxygen isotope has an extra neutron, which appears as one of the final products with, again, the most abundant isotope of neon.

The reactions given here involving α-particle bombardment are typical of a large number of such reactions that can take place.

17.3 Reactions with Protons

Proton bombardment of atomic nuclei can give a γ ray, neutron or α-particle in the final products. One of each kind of reaction is now given;

$$^{1}_{1}p + ^{30}_{14}Si \rightarrow ^{31}_{15}P + \gamma,$$

in this reaction proton bombardment of the least common stable isotope of silicon (3% of silicon) gives the only stable isotope of phosphorus plus a gamma photon.

$$^{1}_{1}p + ^{26}_{12}Mg \rightarrow ^{1}_{0}n + ^{26}_{13}Al,$$

here the bombardment of the second most common isotope of magnesium (11% of magnesium) gives a neutron plus a radioactive isotope of aluminium. This aluminium isotope, with a half-life of 717 000 years plays an important role in some astronomical contexts.

$$^{1}_{1}p + ^{16}_{8}O \rightarrow ^{4}_{2}He + ^{13}_{7}N.$$

In this reaction the final product is an α particle plus a radioactive isotope of nitrogen with a half-life of 10 minutes.

17.4 γ-ray Induced Reactions

Reactions induced by γ rays are not very common and involve the break-up of the impacted nucleus. One such example is

$$^{9}_{4}Be + \gamma \rightarrow ^{1}_{0}n + ^{8}_{4}Be.$$

Beryllium-9 is the only stable isotope of that element. Beryllium-8 has a very short half-life but is an important intermediate component in the way that, at a certain stage in the life of some stars, carbon-12 can be made by the combination of three α particles — the so-called *triple-α process*. The first stage of this process is the combination of two α particles to give the unstable beryllium-8

$$^4_2He + {}^4_2He \rightarrow {}^8_4Be.$$

The beryllium-8 has a very fleeting existence (half-life 7×10^{-17} seconds) but at a late stage of a star's existence its core collapses, thus producing a very high temperature so that α particles come together at a very high rate. Under these circumstances enough beryllium-8 is constantly present to allow another reaction to take place, which is

$$^4_2He + {}^8_4Be \rightarrow {}^{12}_6C + \gamma \quad (Q = 7.4 \text{ MeV}).$$

The energy produced in this reaction raises the temperature of the core even further and the process continues at an increasing rate until the core is almost completely converted to carbon.

17.5 Neutron Induced Reactions

Neutrons are extremely effective in producing nuclear reactions. Because they have no charge they do not experience repulsion due to the positive charge of the nucleus — as happens with α particles and protons. The three main types of neutron-induced reactions give a γ photon, a proton and an α particle respectively as one of the reaction products. An example of each of these types of reaction follows.

$$^1_0n + {}^{12}_6C \rightarrow {}^{13}_6C + \gamma.$$

This reaction converts the most common stable carbon isotope to the less common one (1.1% of carbon).

$$^1_0n + {}^{33}_{16}S \rightarrow {}^1_1p + {}^{33}_{15}P.$$

The bombarded sulphur nucleus is a minor stable isotope (0.75% of sulphur) and the final phosphorus isotope is unstable with a half-life of 25.3 days.

$$^1_0\text{n} + ^{21}_{10}\text{Ne} \rightarrow ^4_2\text{He} + ^{18}_8\text{O}.$$

Neon-21 is a minor stable isotope forming 0.27% of neon while oxygen-18 is also stable and is 0.2% of normal oxygen.

17.6 Fission Reactions

We have noted that neutrons are particularly effective in inducing nuclear reactions because they have no electric charge. From a rather simplistic point of view it might be thought that reactions would be more likely to take place if the neutrons are very energetic, that is to say are moving quickly, but it turns out that this is not so. In general slow neutrons are far more effective in inducing nuclear reactions. The Italian physicist, Enrico Fermi (Figure 17.1; Nobel Prize for Physics, 1938) was the first to discover that this was so.

Fermi carried out a series of experiments in which he bombarded various elements with neutrons and he noted that, as expected, the products of these reactions were elements close in the Periodic Table to the bombarded element. He found that if the neutrons were passed

Figure 17.1   Enrico Fermi (1901–1954).

through paraffin wax before they struck the target the number of reactions was greatly enhanced. He was able to determine this because, if the product element was radioactive, then the fact that more had been produced could be detected by the increased radioactivity. The effect of the paraffin wax was that, in passing through it, the neutrons interacted with many protons (hydrogen nuclei), particles of similar mass, and through these interactions they lost much of their original energy. The slower neutrons spent longer in the vicinity of the target nuclei and so had a greater probability of reacting with them.

One of the elements Fermi bombarded was uranium but the product of that particular reaction was misinterpreted; Fermi thought that he had produced a new non-natural element of atomic number higher than that of uranium but actually he had achieved something of much greater significance, without knowing that he had done so.

In 1938 Otto Hahn and Lise Meitner (Figure 17.2), the pair who had measured β-particle energies (§16.4), together with an assistant Fritz Strassman, were working in the research field pioneered by Fermi and were bombarding uranium atoms with slow neutrons. Before the results of this experiment were properly investigated and

Figure 17.2 Lise Meitner and Otto Hahn.

analysed, Meitner, who was Jewish, left Germany, where at that time anti-Jewish laws were in operation, and with the help of Hahn and two Dutch scientists emigrated to Sweden after illegally crossing the German border into Holland. Hahn and Meitner still collaborated through correspondence with each other and by holding a clandestine meeting in Copenhagen. Hahn and Strassman continued the work on bombarding uranium with neutrons and, a few months after Meitner had left, they detected barium as a product of the reaction. Hahn concluded that uranium had been broken up by the bombardment but he did not understand how this had happened, as he stated when he communicated his results to Meitner. At the time his letter arrived Meitner was being visited by her nephew, Otto Frisch (Figure 17.3), an Austrian refugee working in Copenhagen with Neils Bohr, and they concluded that *atomic fission* had taken place with uranium nuclei splitting up into smaller nuclei, probably barium and krypton. They gave their explanation in a scientific paper in which they estimated the amount of energy that would be released in such a process. The products of fission have less mass than the original uranium atom and the lost mass would appear as energy according to Einstein's equation linking mass and energy. When Frisch returned to Copenhagen he

Figure 17.3 Otto Frisch (1904–1979).

confirmed Hahn's experimental results and isolated the products of the fission process.

17.7 The Atomic Bomb

In 1939 Frisch was on a scientific visit to Birmingham in the UK but was prevented from returning to Copenhagen by the outbreak of World War II. He stayed in Britain and began collaborating with a fellow German refugee and physicist Rudolf Peierls (1907–1995). Together they produced a document known as the *Frisch–Peierls memorandum* that described how an explosive device could be made using uranium-235, the minor component at the 0.7% level, of natural uranium. The basic mechanism of the device is quite simple and is illustrated in Figure 17.4. A slow neutron adds itself to a uranium-235 nucleus to produce uranium-236 that is highly unstable and breaks up into two parts, here shown as barium-144 and krypton-89 although other modes of break-up are possible. Other products of the fission are three high-energy neutrons but if there is some material, such as graphite, around that can act as a *moderator* and remove energy from

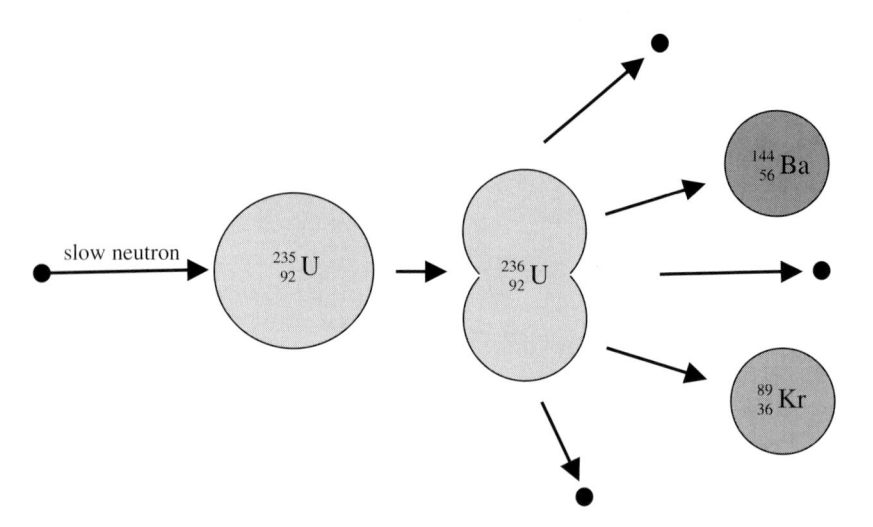

Figure 17.4 The fission of uranium-235 to give barium and krypton nuclei plus three neutrons.

the neutrons then these neutrons can more efficiently cause the fission of other uranium-235 nuclei. It can be seen that this is a runaway process with a rapidly increasing rate of fission events taking place and each fission produces energy. If there is very little uranium-235 present then some of the three neutrons, or perhaps even all of them, will escape without causing further fission and if on average each fission produces less than one neutron available to produce another fission then there will be no explosion, just a rather low rate of energy production. However, with more than some critical mass of uranium-235, calculated by Frisch and Peierls to be about half a kilogram but actually of the order of a few kilograms, the number of effective neutrons per fission will be greater than one and there will be an explosion. In an actual bomb two or more sub-critical masses are kept apart and then brought together to detonate the bomb.

The *Frisch–Peierls memorandum* was a confidential document, sent via influential British scientists to the British government, and it led to the setting up of a project, code named the Tube Alloys Project to look into the possibility of creating an atomic bomb. At the same time there was interest in the United States in this topic. In 1939 a letter had been sent by Albert Einstein and a Hungarian, later American, physicist Leó Szilárd (1878–1969) to the American President, Theodore Roosevelt, warning him that the Germans might be working on atomic weapons and urging him to initiate an American project to do the same. Work did go on at rather scattered sites without much coordination but when America entered World War II in December 1941 this work was drawn together, and became the Manhattan Project, which combined both the American and British activities.

In 1938 Fermi, who had a Jewish wife, emigrated from Italy to the United States to avoid the anti-Jewish measures that were being introduced in Italy. He first went to Columbia University and then to the University of Chicago. Although it was necessary to have uranium-235 to produce an atomic bomb of practical mass for delivery, a sufficient mass of natural uranium in a compact volume is able to sustain a nuclear chain reaction. In December 1942, Fermi and his colleagues constructed the world's first nuclear reactor called Chicago

Pile-1. This consisted of layers of material consisting in total of 400 tons of graphite, acting as a moderator, 58 tons of uranium oxide and 6 tons of uranium metal. To control the rate of production of slow neutrons the reactor was provided with *control rods* covered with cadmium, a material that readily absorbs neutrons. Boron is another element that can be used in this way. If the output of the reactor rose above some predetermined level then the rods were inserted to a greater extent to mop up some of the neutrons being produced and if the output fell then the control rods were partially withdrawn. By modern standards this reactor was primitive in the extreme — for example, there was no external shielding against escaping neutrons — but it was the prototype for building a massive reactor at Hanford, Washington as part of the Manhattan Project. The Hanford reactor was also used for making plutonium, an element of atomic number 94 — beyond the range of the original Periodic Table. This is a fissile material like uranium-235 and it is suitable for bomb making. A layer of natural uranium, mostly uranium-238 was set up round the central core of the Hanford reactor. Neutrons coming from the core reacted with uranium-238 as follows:

$$^{1}_{0}n + ^{238}_{91}U \rightarrow ^{239}_{92}U.$$

The uranium-239 has a half-life of 23.5 minutes, decaying by

$$^{239}_{92}U \rightarrow ^{239}_{93}Np + e^-$$

to give unstable neptunium-239 with a half-life of 2.35 days. The neptunium then decays by

$$^{239}_{93}Ne \rightarrow ^{239}_{94}Pu + e^-,$$

the fissile plutonium having a half-life of 2.44×10^4 years. Plutonium was easily separated from the uranium by chemical means so obtaining plutonium was much easier than obtaining pure uranium-235. To obtain uranium-235 the uranium must first be converted into a gaseous form as uranium hexafluoride, UF_6. This gas is then placed in a centrifuge, a device that spins at extremely high speed thus creating huge g-forces on the material. Uranium hexafluoride molecules containing uranium-235 are less massive than those containing uranium-238 and so less readily move outwards from the spin axis. In this way there is a slight separation of the uranium isotopes but by removing the uranium-235-enriched component of the gas and then re-centrifuging the separation can be steadily increased. By the use of cascades of centrifuges almost complete separation is possible.

The atomic bomb was eventually developed and was tested in the Nevada desert in July 1945. A uranium-235 bomb was dropped on the Japanese city of Hiroshima on 6 August 1945 and a plutonium bomb on Nagasaki on 9 August 1945 — which almost immediately led to the end of World War II. The dropping of these two bombs, directly or indirectly, led to over 200 000 deaths. Since that time there have been about 1 000 tests of atomic bombs — some based on uranium as the fissile material, others on plutonium and yet others were thermonuclear weapons using fusion reactions — known as hydrogen bombs (§17.9). Fortunately no atomic weapons have been used in anger since those dropped on the Japanese cities.

17.8 Atomic Power

The first use of fission reactions was completely destructive but soon thoughts turned towards more constructive uses. What was required for a useful power station was a more-or-less steady output of energy at a rate of hundreds of megawatts. The reactor Chicago Pile-1 was a prototype for future power reactors — although its power output was less than 1 watt! By the use of partial enrichment of the uranium the critical masses for a chain reaction could be reduced. For a steady output of energy the requirement is that each fission event should, on average, produce just one slow neutron causing another fission event. In its simplest form a nuclear power station could consist of a large vessel containing a mixture of enriched uranium mixed with a moderator. Running through the vessel there would be tubes containing a fluid that would take up the heat being generated in the reactor, eventually to produce superheated steam that would operate the turbines that generated electricity. In addition there would need to be control rods to keep the production of energy within required limits.

This is a much-simplified description of a nuclear power station. There are many designs of nuclear power stations and usually the uranium fuel is enclosed in rod-like containers. Again this description glosses over very challenging engineering problems that have to be solved, not only for the efficient running of the power station but also its safety. There have been numbers of very minor accidents at nuclear power stations all over the world, for the most part not well publicized. However, there have been two major incidents that are very well known. On 28 March 1979 a reactor at Three Mile Island in Pennsylvania lost its cooling water and the resulting meltdown of the reactor led to a massive radiation leak. This required the evacuation of 200 000 people from the vicinity of the reactor. On a much greater scale was the explosion on 26 April 1986 at the Chernobyl reactor in the Ukraine, then part of the Soviet Union. A radioactive cloud contaminated large parts of eastern, western and northern Europe and even northerly parts of North America. The incident required the resettlement of more than 300 000 people and is estimated to have led to about 10 000 deaths over the course of time due to the effects of radiation.

Nuclear power was introduced with the optimistic view that it would provide electricity at very low cost — indeed, some supporters asserted that the bulk of the cost of electricity would be that of its transmission. Later it was realized that there were delayed costs of the disposal of the radioactive waste and the problem of very long-term safe storage has still not been completely solved. Over time in many countries there developed an increasing resistance to building more reactors but this now seems to be declining since nuclear energy is seen as a way of producing power without also producing large amounts of the greenhouse gasses that are leading to global warming.

17.9 Fusion — Better Power Production and Bigger Bombs

Power production by nuclear reactions happens in stars. The overall effect of the reactions that take place is that four atoms of hydrogen come together to produce a helium atom with a loss of mass that is converted into energy. The process actually involves three steps

Step 1 $\quad {}^1_1p + {}^1_1p \rightarrow {}^2_1D + e^+ + \nu \quad$ (Q = 1.442 MeV).

In this step two protons come together to give a deuterium nucleus, a positron and a neutrino.

Step 2 $\quad {}^2_1D + {}^1_1p \rightarrow {}^3_2He + \gamma \quad$ (Q = 5.493 MeV).

Here a deuterium nucleus and a proton combine to give a helium-3 nucleus, a stable isotope of helium plus a gamma photon.

Step 3 $\quad {}^3_2He + {}^3_2He \rightarrow {}^4_2He + 2{}^1_1p \quad$ (Q = 12.859 MeV).

Two step-1 processes plus two step-2 processes plus one step-3 process is equivalent to four protons becoming one helium nucleus with the release of 26.73 MeV.

This is known as a *fusion* process because it combines smaller nuclei together to produce a larger nucleus. While it happens within the Sun to provide the energy it radiates it is, in fact, quite inefficient in terms of the power per unit mass. The power output from the Sun is at a rate of 4×10^{26} watts and its mass is 2×10^{30} kilograms so that the power output per unit mass is 0.000 2 watts per kilogram. By contrast, the heat output of a human body with mass 70 kilogram is about 70 watts, or 1 watt per kilogram — all produced by chemical reactions within the body.

Fusion is an attractive process for power production because its products are not radioactive but the solar process, which requires hydrogen at a huge density and a sustained temperature of tens of millions of degrees, is clearly not a practicable proposition to be carried out on Earth. The alternative fusion process that is being considered is one involving two isotopes of hydrogen, deuterium that occurs naturally and tritium, ${}^3_1T$, a radioactive isotope of hydrogen with a half-life of 12.3 years that can be produced in a nuclear reactor by irradiating lithium-6 with neutrons. The power-producing reaction is

$$^2_1D + {}^3_1T \rightarrow {}^5_2He \rightarrow {}^4_2He + {}^1_0n \qquad (Q = 18 \text{ MeV}).$$

The deuterium and tritium nuclei combine to give an unstable nucleus of helium, helium-5, which quickly breaks down into stable helium-4 plus a neutron.

In order for this reaction to happen a temperature of about 100 million degrees is required. This fusion reaction is the basis of a hydrogen bomb, mentioned in §17.7. A mixture of deuterium and tritium is surrounded by a conventional atomic bomb that, when it explodes, creates the conditions for fusion reactions to take place. It is usual to express the power of bombs in terms of the equivalent amount of exploding TNT required to give the same amount of energy. The Hiroshima bomb gave the equivalent of 20 kilotons of TNT, but the largest hydrogen bomb so far tested was the equivalent of 50 megatons — equivalent to 2 500 Hiroshima bombs!

Reaching the required temperature for fusion in more peaceful applications can be achieved by passing a huge electric current through a mixture of deuterium and tritium in the form of a plasma — essentially a gas with electrons stripped off the atoms, so producing a conducting medium containing positively charged ions and negatively charged electrons. The gas can be heated in a pulsed mode but there is then the problem of containing a gas at such a high temperature. This is achieved by the use of what is called *magnetic confinement*. The plasma is contained in a toroidal-shaped vessel, something like a lifebelt, and is constrained by a magnetic field to move round along the axis of the toroid keeping away from its walls.

There are several projects in various countries attempting to develop a practical process of fusion power production. If the quest to produce fusion power in useful quantities is successful then it would represent a great benefit to humankind — but it is not an immediate prospect.

Chapter 18

Observing Matter on a Small Scale

18.1 Seeing

Of all the senses sight is the one that must be regarded as supreme. It has the greatest range of the senses and can operate through the vacuum of space so that we can see stars and galaxies, at distances so great that the light has taken millions of years to reach the Earth. Not only can we see very distant objects, but we can also see what is near to us; a buttercup, seen at a distance as a yellow blob with thousands of fellow flowers in a field, turns out to be an object of intricate beauty when seen at close quarters (Figure 18.1).

The closer we look at the buttercup the more detail we see but there is a limit to the fineness of detail we can resolve with our eyes. This is controlled by the visual *near point*, the closest distance that the eye can focus on an object to produce a clear image.

Forming a visual image is a multi-stage process, illustrated in Figure 18.2. We consider two points on the object, A and B, to illustrate how the image of the whole object is formed. Light, a form of electromagnetic radiation (§9.1), falling on the two points, is scattered in all directions and some of this light enters the eye. The lens of the eye then focuses the scattered rays onto the retina of the eye where photoreceptors transform the light energy into electric impulses. These are transmitted via the optic nerve to the visual cortex of the brain where they create the image. As will be seen, the

170   *Materials, Matter and Particles*

Figure 18.1   A typical buttercup (sannse, Great Holland Pits, Essex).

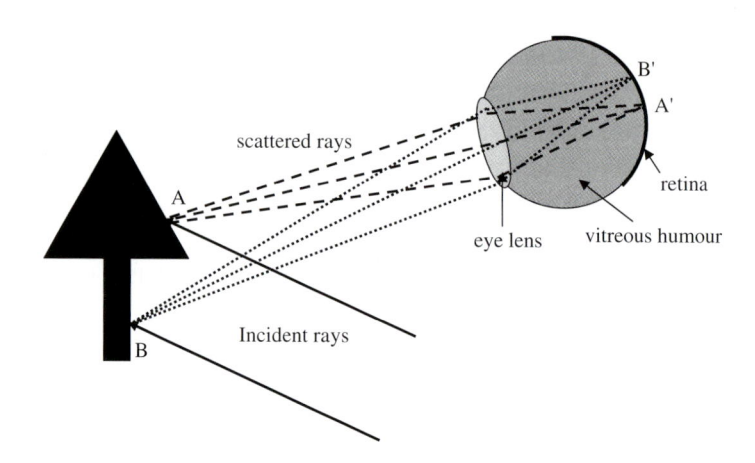

Figure 18.2   The process of forming a visual image.

image on the retina is upside-down but the brain resolves this problem and presents us with a more-or-less accurate impression of the object.

There are important conditions that must be satisfied to produce a sharp image. The electromagnetic wave that represents the light has peaks and troughs that move along with the speed of light, which varies with the medium through which the light moves. When a light ray falls on the point A all the scattered rays from that point that move towards the eye all start off *in phase*, which means that at the point A their peaks and troughs occur at the same time. The rays then travel along different paths, including their passage through the lens of the eye and the *vitreous humour*, the fluid occupying the space between the lens and the retina. For a sharp image of the point A to be produced, when the different rays fall on the retina they must all arrive in phase, which is with peaks and troughs together. This requires that the time of travel by the different scattered rays from point A to point A' should all be the same. Another condition is that the object which is to be seen should not be too close to the eye, for then the lens is unable to adjust to the extent where it can bend all the scattered rays to come together at one point. As a general rule, the closer the object is the more detail can be seen but once the object comes closer than the near point that rule breaks down.

To be able to see more detail in an object than is possible just by the human eye alone one must resort to optical aids.

18.2 Microscopes

It is a matter of everyday experience that it is possible to see more detail in an object by the use of a simple lens, often known as a *magnifying glass*. The way that this works is illustrated in Figure 18.3. The rays from point A of the object are bent by the lens and appear to have diverged from point A' and, similarly, the rays from B appear to have come from B'. Now the image A'B' has the same *angular size* as seen from the centre of the lens as the original object AB and, since the eye is close to the lens, this condition will also be approximately satisfied for the eye. The advantage factor of using the magnifying glass is that the object AB is closer to the eye than the near point, and hence could not be clearly seen, while the apparent object A'B', which seems to be the source of the scattered rays, can be at the near point and hence be seen in sharp focus.

There are physical constraints on the size of a magnifying glass for normal everyday use and hence a limit to the magnification that can be achieved although, under special conditions, quite remarkable

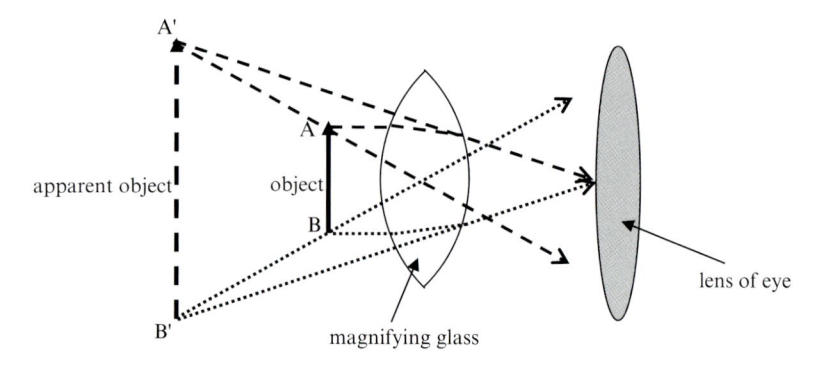

Figure 18.3 The action of a magnifying glass.

Figure 18.4 Hans Lippershey (1570–1619).

magnifications can be obtained. A magnifying glass is sometimes called a *simple microscope*; the 'simple' part of the name refers to the fact that just a single lens is used. Much greater magnifications can come about by the use of a *compound microscope*, usually just called a *microscope*, a device containing two or more lenses. The inventor of the microscope is not known for certain but the invention is usually credited to Hans Lippershey (Figure 18.4) a German lens-maker who

Figure 18.5 Hooke's microscope.

spent most of his life in Holland. He was also responsible for designing the first practical telescope, later copied and used for important astronomical observations by Galileo Galilei (1564–1642).

The essentials of the modern microscope were present in the one designed and used by Robert Hooke (1635–1703), a prominent English scientist who was in frequent conflict with Isaac Newton on many scientific issues of the day. Hooke's microscope is illustrated in Figure 18.5. The specimen to be examined is illuminated by a source of light, often concentrated by a lens, and is observed through a tube containing at least two lenses, the one closest to the specimen being termed the *objective* and the one closest to the eye being called the *eyepiece*.

Hooke's observations through his microscope, as shown by drawings of the images he saw, were a revelation of a world beyond the visual range of the unaided eye or even the magnifying glass. His drawing of the image of a flea (Figure 18.6) was typical of the detail shown in many other biological systems.

No matter how perfect is the optical system of a microscope there is a limit to the resolution that can be achieved that depends on the

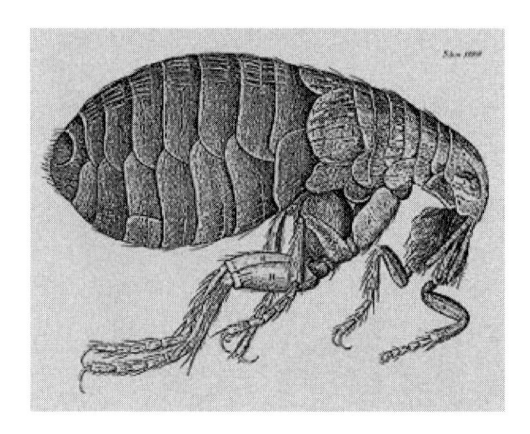

Figure 18.6 Robert Hooke's drawing of a flea.

wavelength of the radiation used to do the imaging. As a general rule the least distance that can be resolved in an image is approximately equal to the wavelength that, for visible light, is between 0.4 μm[h] at the blue end of the spectrum to 0.7 μm at the red end. One way of increasing the resolution is to go down further in the electromagnetic spectrum and to use shorter wavelengths in the ultraviolet region. This involves the use of special optical materials that are transparent to ultraviolet light, such as quartz or the mineral fluorite, and the transformation of the image into a form that can be seen, but these technical problems can be solved and ultraviolet microscopy is a well-established technique.

There are limits of wavelength, and hence of resolution, that can be reached with electromagnetic radiation in the context of microscopy. The maximum magnification that can be achieved with visible light or ultraviolet radiation is about 2 000. To get to magnifications of 1 000 000 or more would require wavelengths equivalent to X-rays but although there is no problem in producing X-rays what cannot be done is to focus them in the way that a lens does for visible light. For this reason, to attain higher resolution one must resort to electron microscopy using the fact that high-energy electrons can

---

[h] 1 μm is 1 micrometre = 10^{-6} metre.

behave like waves in the way described by Louis de Broglie (§11.2). Electrons accelerated through a potential difference of 1 000 volts have an equivalent wavelength of about 4×10^{-11} metre, some 10 000 times less than the wavelength of blue light. The focusing of the electrons scattered from the specimen to be examined is done with electric and magnetic fields, which do not have the precision of optical lenses so instead of a 10 000-fold advantage in magnifying power the actual advantage is closer to 1 000. Even so, electron microscopes are capable of giving magnifications of 2 000 000, or even better. Figure 18.7 is an *electron micrograph* (a picture taken with an electron microscope) of helical carbon nanotubes, a form of carbon with structure on a tiny scale. The bar at the bottom left of the picture is 50 nanometres long, just one-tenth of the wavelength of green light. Carbon in this form has different electrical properties to those of bulk carbon and it has a number of potential technical applications (§22.4).

The helical carbon nanotube was imaged by *transmission electron microscopy*, which can be used when the specimen is thin and is transparent to the electron beam. Another form of electron microscopy is *scanning electron microscopy* in which the electron beam makes a raster scan over the object and picks up the back-scattered electrons. Depending on what type of detector is used the information can be a profile of the object or some other property such as the variation

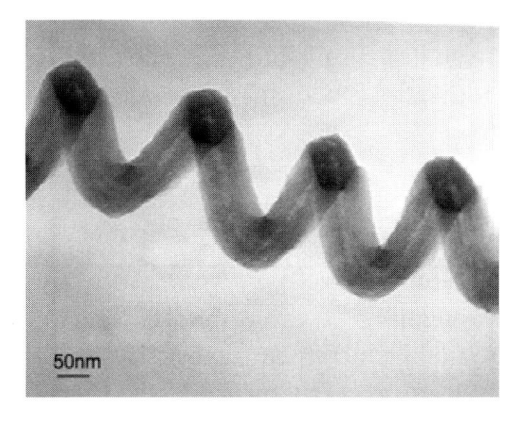

Figure 18.7 A helical carbon nanotube (UC San Diego Jacob School of Engineering).

176   *Materials, Matter and Particles*

Figure 18.8   An electron micrograph of pollen grains (Dartmouth Electron Microscope Facility. Dartmouth College).

of chemical composition over its surface. Figure 18.8 shows a scanning electron micrograph of pollen grains, the sizes of which vary from about 10^{-5} to 10^{-4} metre. The fine detail seen in this picture corresponds to a resolution of 10^{-7} metre or better, where optical microscopy would be straining at its limit.

Electron microscopy can produce images with a best resolution corresponding to about ten times the distance between the atoms that constitute matter. However, for many purposes it is required to explore the structure of matter at even higher resolution and actually to image individual atoms in some way. This cannot be done by conventional imaging but requires the use of a technique that has revolutionized science since its discovery a century ago — X-ray diffraction.

18.3 X-ray Diffraction from Crystals

By 1912 it was accepted that X-rays were a form of electromagnetic radiation with wavelengths of the order 10^{-10} metres. As previously noted, an important characteristic of light that is due to its wave nature is that it can undergo the process of diffraction and one way of showing this is with a diffraction grating — a series of uniformly and finely spaced transmitting slits separated by opaque regions.

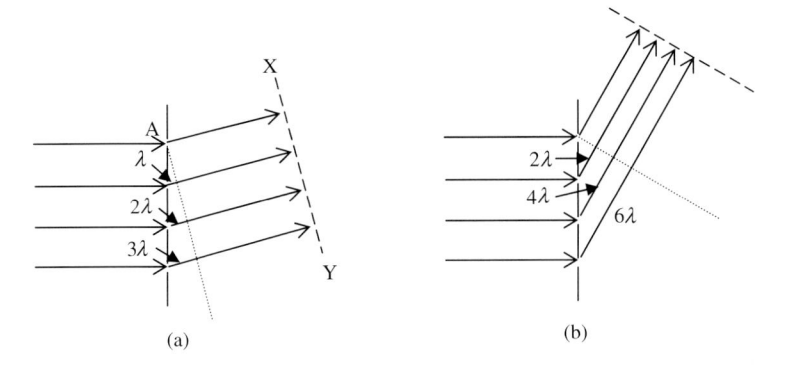

Figure 18.9 The formation of two diffracted beams by a diffraction grating.

A representation of a diffraction grating and the way that it acts on a beam of light is shown in Figure 18.9. When light falls on a clear slit it is scattered from the slit in all directions. The different slits receive the crests and troughs of the wave motion of the oncoming radiation in unison. Now we consider the light scattered in the direction shown in Figure 18.9(a). The light coming from the slit below A in that direction is delayed by one wavelength (λ) behind that coming from A. Since crests (and troughs) of the wave motion are separated from each other by one wavelength the rays from A and from the slit below are in unison crossing the dashed line XY and the same applies to rays coming from all the other slits. Thus the rays reinforce each other and a strong *diffracted beam* is formed in that direction. Figure 18.9(b) shows another direction for a strong beam where, in this case the rays from neighbouring slits have a path difference of two wavelengths (2λ). The overall effect of a diffraction grating is to produce a set of diffracted beams in various directions and the condition for it to produce this effect is that the separation of the slits has to be greater than, but of the same order of magnitude as, one wavelength.

In 1912 the German physicist Max von Laue (Figure 18.10) was taking a stroll in the Englischer Garten in Munich with a young student, Paul Ewald (1988–1985), who was describing the topic of his doctoral thesis concerned with the interactions of light with a crystal. It then occurred to von Laue that since a crystal was a

Figure 18.10   Max von Laue (1879–1960).

periodic three-dimensional object, an arrangement of atoms repeated on a regular lattice in three dimensions with repeat distances larger than, but comparable to, the wavelengths of X-rays, then crystals might form a three-dimensional diffraction grating for X-rays. He asked two researchers in his laboratory, Walter Friedrich (1883–1968) and Paul Knipping (1883–1935), to fire a beam of X-rays at a crystal to see what the result would be. What they found on a photographic plate placed on the opposite side of the crystal from the oncoming beam was a pattern of spots confirming that diffraction had indeed taken place. One of their X-ray diffraction photographs, for the crystalline mineral zincblende, is shown in Figure 18.11. Von Laue received the 1914 Nobel Prize for Physics for this discovery. It was one that was to have enormous repercussions for many fields of science.

When the paper by von Laue, Friedrich and Knipping was published in 1912 it was accompanied by an interpretation of the results that described the diffraction spots as being due to five different wavelengths in the X-ray beam. This paper created quite a stir in the scientific community and aroused great interest in a young Australian-born British physicist William Lawrence Bragg, known as Lawrence Bragg, (Figure 18.12a) who was in his first year as a research student at

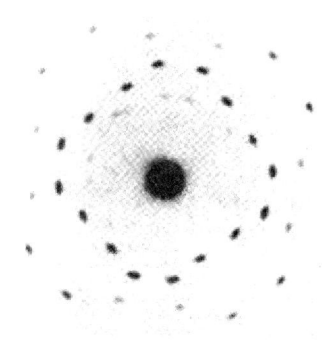

Figure 18.11 Friedrich and Knipping's X-ray diffraction photo for zincblende.

Figure 18.12 (a) William Lawrence Bragg (1890–1971) (b) William Henry Bragg (1862–1942).

the University of Cambridge. The theory in the Laue *et al.* paper neither completely nor correctly explained the diffraction spots but young Bragg deduced that there was a way of describing the diffraction phenomenon in terms of mirror-like reflections from richly populated planes of atoms within the crystal. A two-dimensional illustration of this is given in Figure 18.13. The lines shown are equally spaced and go through all the purple atoms. Similar sets of lines, with similar spacing but displaced, go through each of the other types of atom in the repeated structure. The interpretation of X-ray diffraction by what

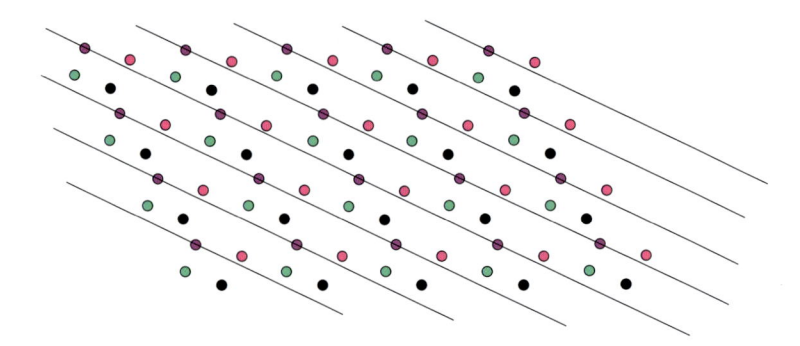

Figure 18.13 A hypothetical two-dimensional crystal showing richly populated lines.

is now known as *Bragg reflection*, which completely explained what was observed, was published in a letter to the journal *Nature* in 1912.

Lawrence Bragg's father, William Henry Bragg (Figure 18.12b) was Professor of Physics at Leeds University and had developed an X-ray spectrometer, a device by which the intensities and wavelengths of X-rays could be accurately measured. This was the device that had been used by Moseley to find the wavelengths of characteristic X-radiation (§10.2). It could also be used to record the intensities of X-ray diffraction beams much more accurately than by the use of photographic plates and the father and son team worked together measuring diffraction patterns for many simple crystals. Most importantly they developed a technique, using the directions and intensities of the X-ray diffracted beams, for finding the arrangement of atoms in the crystals. This was a most important advance; Figure 18.14 illustrates their determination of the structure of sodium chloride, common salt. It consists of a regular lattice of alternating sodium and chlorine atoms in three dimensions and showed that there was no such entity as a *molecule* of sodium chloride. It was a cubical framework structure, bound together by ionic bonding (§13.2), with each atom having six neighbouring atoms of the other variety. Other structures solved were the minerals zincblende (zinc sulphide, ZnS), fluorspar (calcium fluoride, CaF_2), iron pyrites (iron sulphide, FeS_2) and calcite (calcium carbonate, $CaCO_3$). These developments by von Laue and the Braggs

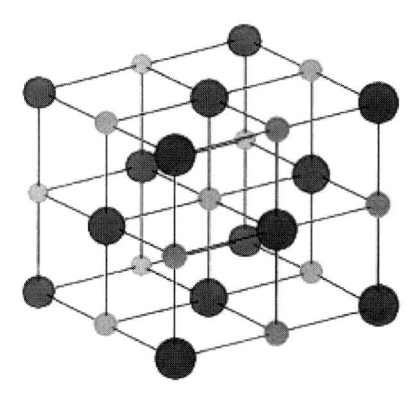

Figure 18.14 The structure of sodium chloride. Large circle = sodium, small circle = chlorine.

were the first steps in what would turn out to be possibly the most important scientific advance of the 20th century.

For their work as pioneers of X-ray crystallography the Braggs, father and son, were jointly awarded the Nobel Prize for Chemistry in 1915. At the age of 25 Lawrence Bragg was the youngest ever recipient of the prize and he received news of the award while he was serving as an artillery officer in France during World War I. Paul Ewald, whose thesis work had been influential in turning von Laue's mind towards the idea of X-ray diffraction, was serving in the German army on the other side of the front line, manning an X-ray radiography unit to help in the treatment of wounded German soldiers. Ewald had one Jewish grandparent and a Jewish wife so in 1937 he emigrated from Germany, at that time governed by an anti-Semitic National Socialist government, and settled in England. In 1944 Ewald proposed the establishment of the International Union of Crystallography, the body that now oversees that subject on a worldwide basis, and in this enterprise he was greatly helped by Lawrence Bragg with whom he had established a strong friendship.

Chapter 19

Living Matter

19.1 Defining Life

We have an instinctive understanding of what constitutes life. When an ant is seen scurrying about in the garden, perhaps carrying a portion of a leaf, there is no doubt that what is being seen is a living creature. When we see a stone in the garden we are equally certain that it is inanimate; it doesn't move and it doesn't change with time, except perhaps by external agencies such as wind and rain over extremely long periods of time. Other entities do not move — for example, trees — but they constantly change with the seasons and they first grow to maturity from an embryonic form and eventually decay and lose their identity. Trees are certainly a form of life — although not animal life.

Sometimes instinct can be misleading. Anyone who saw coral for the first time might be excused for thinking of it as some form of rock, and hence inanimate, but such a conclusion would be wrong. A picture of a typical coral is shown in Figure 19.1. Each pillar-like structure consists of a large number of millimetre-size *polyps*, living organisms that can breed, either sexually or asexually, and take in food in the form of algae or plankton, sometimes using stinging cells attached to fine tentacles to catch their prey.

When it comes to borderline situations not all scientists agree about what constitutes life. However, there are some characteristics that must be present in any definition, which we now list.

Figure 19.1   Pillar coral (Florida Keys National Marine Sanctuary).

Reproduction

Whatever definition of life is taken, at some stage a living organism will die so that it no longer satisfies that definition. The lifespan may be very short by human standards, a few days for a common fruit fly (Figure 19.2a), to very long, nearly 5 000 years for a surviving Bristlecone pine (Figure 19.2b). Since there is no elixir giving eternal life, a necessary condition for the survival of any living form is that it must be able to reproduce. Reproduction can be simple, as by the dissemination of spores, or complex, as that which gives mammalian birth.

Adaptation

Taken over geological timescales, environmental conditions in any location will change and such changes are often, and usually, detrimental to survival of the life forms that exist there. The reaction to such changes may be to migrate to a more favourable locale, but such

Figure 19.2 Examples of extreme lifespans (a) a fruit fly (b) a Bristlecone pine.

a solution may not be available on an island or for plants that do not widely disseminate their seeds. The organism may be able to survive by changing its characteristics in some way. This can come about by gradual changes, following Darwinian principles, whereby those members of the species with characteristics that give it the greatest chance of survival, will be able to pass on those desirable characteristics to their progeny. In this way, over time, the general characteristics of the population will change — adaptation will have taken place.

Sometimes the change to more favourable characteristics can happen in a less gradual way by a process called *mutation* that will be described more fully in the following chapter.

Regeneration and Growth

Complex life forms consist of collections of *cells* that carry out many different functions. Individual cells have a lifetime that is shorter, usually much shorter, than that of the organism of which they form a part. There must be a regenerative process to replace dying cells otherwise the organism itself cannot survive. When the organism first comes into existence it is small and immature and some mechanism is required to produce growth and progress towards maturity. Hence this mechanism must also produce new cells and not just replace former cells that have died.

Metabolism

Organisms must take in food of some sort and convert it either into the proteins and other materials that constitute its substance or into energy that will enable it to carry out essential processes. The chemical reactions that occur in cells to enable this conversion to take place is referred to as *metabolism*.

Response to Stimuli

For survival the organism must be able to respond to stimuli from its surroundings. Even some primitive single-celled life forms (which some may not even define as living) contract when touched. Animals react through many senses linked to brain activity to optimise their survival chances by, for example, avoiding danger or acquiring food. Even vegetable organisms respond by turning their leaves to absorb the greatest possible amount of sunlight.

Not all scientists would agree that the characteristics listed above form a set that is either complete or necessary. Some would think the list to be incomplete while others might feel that it was too prescriptive. In addition there are some entities that are borderline. For example, viruses consist of genetic material enveloped in a protein coat. If they invade a cell they then use the contents of the cell to replicate themselves and so damage or even kill the cell; the living creature of which that cell is a part may then become diseased or even die. Viruses have another life-like characteristic in that they adapt — in fact so quickly and readily that antiviral therapies constantly have to be changed. The vaccine that will prevent an influenza epidemic in one year is normally useless for the influenza virus that follows one year later. Nevertheless, although they have some of the characteristics of living matter, viruses are not usually categorized as a form of life.

19.2 Forms of Life

The range of living matter in terms of its complexity is huge. It stretches from bacteria, which consist of a single cell, to animals, including humankind, which are complex structures consisting of many different components with specialized activities. We start our description of living matter with the simplest forms and work upwards to ourselves.

Bacteria and Archaea

Bacteria are tiny entities, a few microns in extent and consisting of a single cell. They can be of many different shapes — spheres, rods or spirals being the common forms — but, regardless of shape, their basic functions are similar. A representation of a spherical cell is shown in Figure 19.3. The various components are:

Capsule

This is a layer that prevents harmful materials from entering the cell. It is usually a polysaccharide, a type of material of which starch and cellulose are examples. Most, but not all, bacteria possess this feature.

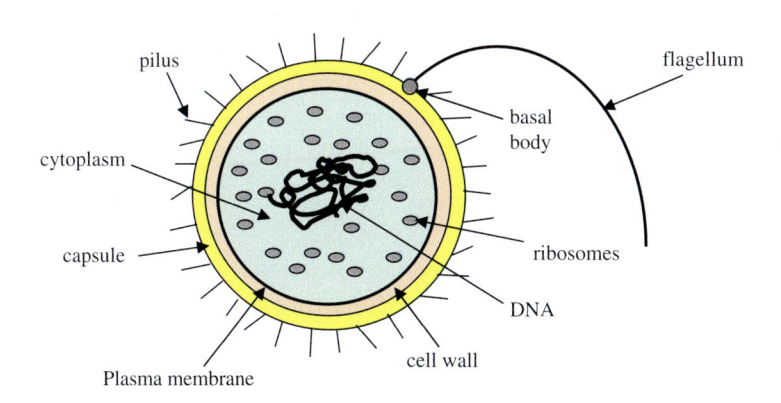

Figure 19.3 Representation of a bacterium.

Cell Wall

This structure defines the shape of the bacterium and it consists of a mixture of proteins and polysaccharides. *Mycoplasma* is a type of bacterium that does not have a cell wall and hence has no definite or permanent shape. Some species of this genus of bacteria are harmful to human health — for example, one that causes pneumonia.

Plasma Membrane

This is the layer through which nutrients are passed into the cell and waste materials leave it. It consists of lipid materials that are soluble in organic solvents and insoluble in water. Human body fat is one form of lipid.

Cytoplasm

This is a jelly-like material within which reside the ribosomes and the genetic material, deoxyribonucleic acid (DNA).

Ribosomes

There are many of these small organelles (components of an organism) in the bacterial cell giving it a granular appearance in an electron micrograph. Ribosomes are the locations where the proteins required by the cell for its survival and development are produced. Within them is a material called mRNA (messenger ribonucleic acid) that controls which proteins are produced.

DNA

DNA is the basic genetic material, the structure of which defines every feature of an organism. It is the information contained in DNA that produces the mRNA needed by the organism to create the proteins it requires.

Pili (*plural of pilus*)

These hair-like hollow structures enable the bacterium to adhere to another cell. They can be used to transfer DNA from one cell to another.

Flagellum and Basal Body

The flagellum is a flail-like appendage that can freely rotate about the basal body, which is a universal joint, and so provide a mode of transport for the bacterium. A single bacterium may have one or several flagella (plural of flagellum).

Bacteria are found in a wide variety of environments — almost everywhere in fact. Figure 19.4 shows a bacterium found in Arctic ice; one wonders how it got there and what it is doing there! There are about ten times as many bacteria residing on and in a human body as there are cells that constitute the body itself, mostly residing on the skin and in the digestive tract. The world population of bacteria is estimated at about 5×10^{30}; if they were all lumped together they would occupy a cube with sides about 10 kilometres long. Another comparison is that the mass of bacteria on Earth exceeds the mass of all humanity on Earth by a factor of between 1 000 and 10 000! Bacteria may be tiny but there are many of them.

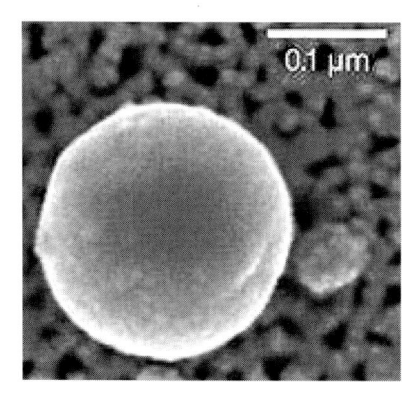

Figure 19.4 A bacterium found in Arctic ice (National Science Foundation).

The popular view of bacteria is that they are harmful and the world would be a better place without them. This is far from the truth. Some of them are dangerous; millions of people die each year from bacterial infections such as tuberculosis and cholera. However, through its immune system the body normally provides us with protection against most bacteria and when it does not do so we can call on antibiotics that deal with many of them. On the positive side, bacteria can be very useful and even necessary to other life forms. They are essential in recycling nutrients; when a tree dies bacteria breaks down its substance so that it can be absorbed in the soil and be taken up by other plants. Bacteria that are present in the roots of some plants carry out the fixation of nitrogen from the atmosphere — the natural formation of plant fertilizers.

In the 1970s scientists were finding bacteria in very hostile environments of extreme temperature, salinity, acidity or alkalinity. Some of the first of these were found in the hot springs of Yellowstone Park in the USA. However, when the genetic structures of these 'bacteria' were examined it was found that they were completely different from bacteria and not simply a bacterial adaptation to extreme conditions. They look like bacteria — they take on similar forms — but they are not bacteria and they are now called *archaea*.

Archaea are found in locations as different as the bottom of the ocean near volcanic vents that heat water to over 100°C to the digestive tracts of cows, termites and some sea creatures, where they are responsible for the emission of methane, a greenhouse gas. They are also found in mud in marshland areas and even in petroleum deposits. Although archaea from different habitats are different from each other they are clearly related and are adaptations from a single source.

Since archaea can live in extreme conditions and do not need oxygen to survive it has been suggested that they were the first life forms to evolve, at a time when the extreme conditions on Earth, without an oxygen atmosphere, could not have supported any other form of life that exists now. If that were so, and if bacteria are so different in genetic structure that they could not have evolved from archaea, then it offers the intriguing possibility that life arose on Earth on two occasions quite independently of each other. There are two extreme assumptions about the origin of life. One is that it is an event so unlikely that it may have happened only once — and hence that we humans are the only intelligent witnesses to its origin. The other assumption is that where suitable conditions exist life will almost inevitably come about — albeit that we have not yet discovered the process by which this would occur. If life did arise twice, and independently, on our planet then it would give strong support to the latter assumption.

Eukaryota

Bacteria and archaea consist of a single cell similar to that shown in Figure 19.3. Eukaryota are organisms that consist of one or more cells, but these cells are complex and contain the genetic material within a nucleus bounded by a membrane. Within this class of life forms there are four *kingdoms* — subgroups of connected distinctive organisms.

Protista

These are single-celled organisms plus some of their very simple multi-celled relatives. Single-celled protista include amoebas and diatoms while slime moulds and various types of algae are examples of the multi-celled variety. An example of this kind of organism is the red algae, shown in Figure 19.5. Red algae often amalgamate into a plant-like form and they are harvested and used as food in some societies.

Fungi

These are a common multi-cellular type of organism of which mushrooms, toadstools, various moulds and yeasts are typical examples. They digest their food externally and then absorb the digested nutrients into their cells; thus many mushrooms are found on rotting fallen trees, which are their source of food. A typical fungus is shown in Figure 19.6. Some mushrooms are edible, while others are poisonous, and yeasts are used both for making bread and fermenting alcoholic beverages.

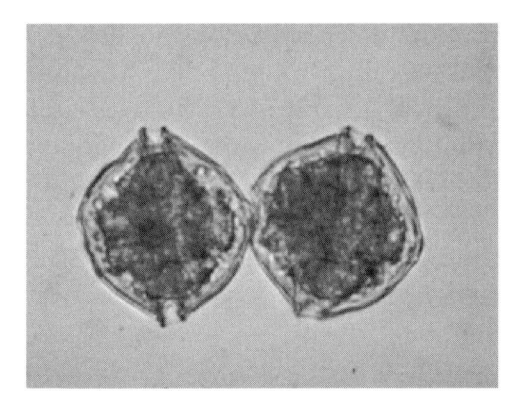

Figure 19.5 Red algae (Woods Hole Oceanographic Institution).

Figure 19.6 A woodland fungus (T. Rhese, Duke University).

Plantae

This kingdom of multi-cellular organisms includes all flowering plants, ferns, bushes and trees. Nearly all of them are green due to the pigment chlorophyll they contain. Chlorophyll uses radiation from the Sun to fuel the manufacture of cellulose, starch and other carbohydrates that make up the structure of the plant. Plants live in a symbiotic relationship to *animalia*, the next kingdom to be described. Chlorophyll takes water plus atmospheric carbon dioxide and from them produces cellulose plus oxygen, which is released into the atmosphere. Thus for plants carbon dioxide is a necessity and oxygen is a waste product. Animals breathe in oxygen, which enables energy to be produced in muscles by chemical processes, and exhale carbon dioxide. For animals oxygen is a necessity and carbon dioxide is a waste product.

Cellulose forms the stems of plants and also the trunks of trees. It is an extremely strong material and the tallest trees are more than 100 metres in height.

Animalia

This is the kingdom to which *homo sapiens* belong. Animals cannot use non-living material as food; they can only ingest the products of life — protista, fungi, plantae or other animals. Animals are the only life form that has two distinct types of tissue — muscle tissue and nervous tissue. Figure 19.7 shows a chimpanzee, the closest relative to man.

In this brief survey of living forms we have dealt with living matter, the material aspect of that which constitutes life. We have seen that

Figure 19.7 Man's closest relative — the chimpanzee (National Library of Medicine).

there is a tremendous range of complexity and animals represent the most complex life form. But life, or at least complex life such as *homo sapiens*, cannot be described just in material terms. This has been known from antiquity. In Chapter 1 we described how the tutor Acacius was teaching that there were not just the four material elements but also a fifth element, a spiritual one. This fifth element, the *quintessence* as it was often called, related to any aspect of the world or its contents that could not be defined in purely materialistic terms. A notable example of this is *consciousness*, an aspect of a human being that seems inexplicable in material terms alone. It gives awareness of our own existence and our role within our environment. It can trigger memories of the past and make plans for the future. It controls feelings such as happiness, sorrow, fear and rage. A bacterium or a tree certainly does not possess the property of consciousness. Can we say the same about a chimpanzee?

Many lowly creatures show behaviour patterns that seem similar to those of man but which may come about for quite different reasons. A soldier ant will sacrifice himself for the good of the colony — but can this be equated to the behaviour of a human soldier who risks, or even gives, his life to save his comrades, or is it a programmed reaction like that of a unicellular organism that recoils to touch? These are difficult questions. Some have tried to explain the action of the human brain, the source of intelligence, consciousness and emotions, in terms of it being a very complicated computer. However, that is not an explanation that many find appealing.

Chapter 20

Life at the Atomic Level

20.1 Seeing Life Matter at the Atomic Level

In §18.1 the basic physics of producing an image was described. The first essential requirement is a source of radiation of a wavelength that is short enough to resolve the details of the object of interest. This radiation is directed towards the object and is scattered in all directions. Some of the scattered radiation is picked up by a lens, or perhaps a combination of lenses, and is brought together to form an image. The lens has the important property that it ensures that radiation scattered in various directions from a point of the object all *in phase*, i.e. with synchronized peaks and troughs, arrives at the image point in phase. Without this condition no image can be formed.

If we wish to look at the arrangement of atoms in a material then the distance to be resolved is the distance between atoms. The appropriate electromagnetic radiation satisfying this condition is X-rays with wavelengths about 10^{-10} metre or, in the units preferred by X-ray crystallographers, 1 Ångstrom unit (Å). It turns out that the majority of the materials important to life, mainly proteins, can form crystals in which the chemical molecules arrange themselves in regular arrays. There is no problem in firing a beam of X-rays at a small crystal and scattering the X-rays from it. X-rays are dominantly scattered by the atomic electrons that form an approximately spherical cloud around each nucleus so an image of an atom formed from the scattered X-rays would look like a cloud of electrons with a concentration of density at the centre where the nucleus is located. However, here the correspondence with forming an optical image must end; no X-ray lens exists so the scattered radiation cannot be brought together to form an image.

In §18.3 we saw that the scattering of X-rays from crystals was of a particularly simple form, consisting of a number of diffracted beams of different intensities in various directions. For a very simple structure, such as sodium chloride or others slightly less simple but simple enough, the Braggs were able to find the atomic structure by various *ad hoc* methods. What they did was to examine the pattern of variation of intensities from one diffracted beam to another and then deduce what arrangements of atoms was necessary to explain it. However, for more complicated structures this approach does not work. Over the period since the Braggs did their pioneer work X-ray crystallographers have devised a number of different ways, some experimental and some mathematical, to find the information that enables them to simulate the action of an X-ray lens by calculating what it would do if it actually existed. These calculations give what are called *electron-density maps* that show the distribution of electron density in one of the repeat units of the crystal structure.

Methods of solving crystal structures have progressed to the point where molecular structures containing many thousands of atoms, such as protein structures, can be solved routinely — a far cry from sodium chloride. Figure 20.1 shows part of an electron-density map for the protein lysozyme. This protein occurs in tears, saliva and egg whites and has the property that it breaks down bacterial walls and so is an anti-bacterial agent. The fuzzy blobs represent the calculated electron density, superimposed on which there is a framework of spheres and rods representing individual atoms and the chemical bonds that link them. This is a projected view of a three-dimensional structure so some groups of atoms forming almost perfect regular hexagons or pentagons are seen in a distorted form. It should be noted that this is only a small part of the structure that contains more than 1000 non-hydrogen atoms. Hydrogen atoms are difficult to

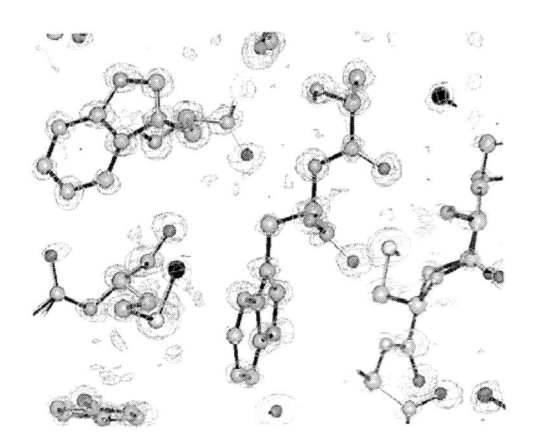

Figure 20.1 A projected view of part of the crystal structure of the protein lysozyme.

detect with X-ray diffraction because, having only a single electron, they scatter X-rays very weakly.

From the middle of the 20th century there have been enormous advances in molecular biology that have led to a detailed understanding of the way that life processes operate at the atomic and molecular level. The outstanding agency for this advance has been X-ray crystallography. Little did the early pioneers of that subject realise the extent of the scientific revolution that they were beginning to create.

20.2 Encoding Complex Structures

We are accustomed to the idea of representing a complex structure as a blueprint, a detailed drawing showing every part of the structure and the way the parts are assembled into the whole. We can have a blueprint for a stately home, a steam locomotive or a suspension bridge and a skilled builder or engineer could work from the blueprint to build the structure as its designer intended. There is a huge variety of shapes and materials that can be incorporated into structures of various kinds. If one wished to find a basic set of components from which all structures, and all blueprints, could be constructed it would be a very large set — indeed one of infinite size.

Another kind of structure is a work of literature, which we will think of as in some European language. From an analysis of a large number of literary works it is found that there are a finite number of words from which all of them could be constructed, say 50 000. In this respect it is interesting to note that in the whole of Shakespeare's huge output of plays and sonnets taken together there are less than 30 000 different words. However, we could break down the whole of, say, English literature into even fewer components — into the 26 letters of the English alphabet. The compositor who wished to set up a Shakespeare play by old-fashioned typesetting would not need 30 000 trays, each holding a different word, but only 26 trays, each holding a different letter. Actually the basic set of components for describing a work of literature could, in principle, be reduced even further. In the 1840s an American, Samuel Morse, devised a scheme for telegraphic communication that depended on a sequence of either short or long signals, known as dots and dashes respectively. In *Morse code* 'a' is '•–', 'e' is '•' and 'p' is '• – – •'. A dash has the length of three dots, the space between the components of the same letter is one dot-length, the space between two letters is three dot-lengths and the space between two words is five dot-lengths. Thus the word 'ape' would be '•.–… •.–.–.•… •'. In this way just three components — a dot, a dash and a space — can be used to transmit the full range and subtlety of a language. In principle a work of literature could be typeset with only three trays, but most people could not read such a coded message and, importantly, it would also occupy much more space.

20.3 Encoding Living Matter — The Chemistry of DNA

We have seen that an important requirement for any species to survive is that it must be able to reproduce itself. In the case of bisexual reproduction the parents must produce a blueprint that will give rise to an infant of the same species. Yorkshire terrier parents will have puppies that are clearly Yorkshire terriers, albeit they will not exactly resemble either parent. Just as a large range of literary material, from a simple message to a major novel, can be constructed from the same 26 letters of the alphabet so it turns out that the basic blueprints for the large range of living forms, from a single-cell bacterium to a human being, can be constructed from a chemical alphabet of just four letters. The key to this coding system is the DNA mentioned in §19.2 and shown as a component of a bacterium in Figure 19.3.

Every feature of an organism, from its appearance, its function and the processes that go on within it, is completely defined by its DNA. The plot of the film *Jurassic Park* (1993) is based on the idea that the DNA of extinct dinosaurs, extracted from the blood ingested by mosquitoes of the Jurassic Period trapped in amber, can be reactivated by inserting it into the eggs of modern birds — the remaining descendants of dinosaurs. The film is science fiction but the basis of it, that any living organism is completely defined by its DNA, is completely valid. DNA is a string of linked chemical units (a *polymer*) of which there are only four basic types; these units, known as nucleotides, are shown in Figure 20.2. The ringed parts are common to all nucleotides and consist of a sugar component plus a phosphate group consisting of a phosphorus atom linked to four oxygen atoms. The difference from one nucleotide to another is the *base* of the nucleotide and is either one of the two *purines* — *guanine* (G) and *adenine* (A) — or one of the two *pyrimidines* — *thymine* (T) and *cytosine* (C). The units are chemically linked by the chemical bond, shown dashed, that joins a sugar in one unit to a phosphate group in the next. The order in which the bases occur defines every aspect of the organism that either produced the DNA or is to be created from it. The complete sequence of letters corresponding to the DNA for a particular organism is its *genome*, which is a set of instructions for constructing that organism. A sequence in part of the genome that gives the instructions for making a particular protein is known as a *gene*, which may involve anything from one thousand to one million bases. The genome is not linked together as one long chain but is arranged in several bundles of genes, each of which is known as a *chromosome*. The term chromosome is of Greek origin and means 'coloured object' because chromosomes readily absorb dyes and can be seen as highly chromatic objects when viewed in a microscope. The human genome contains about three billion bases organized into about 20 000 genes and 46 chromosomes, which are arranged in 23 related pairs. The

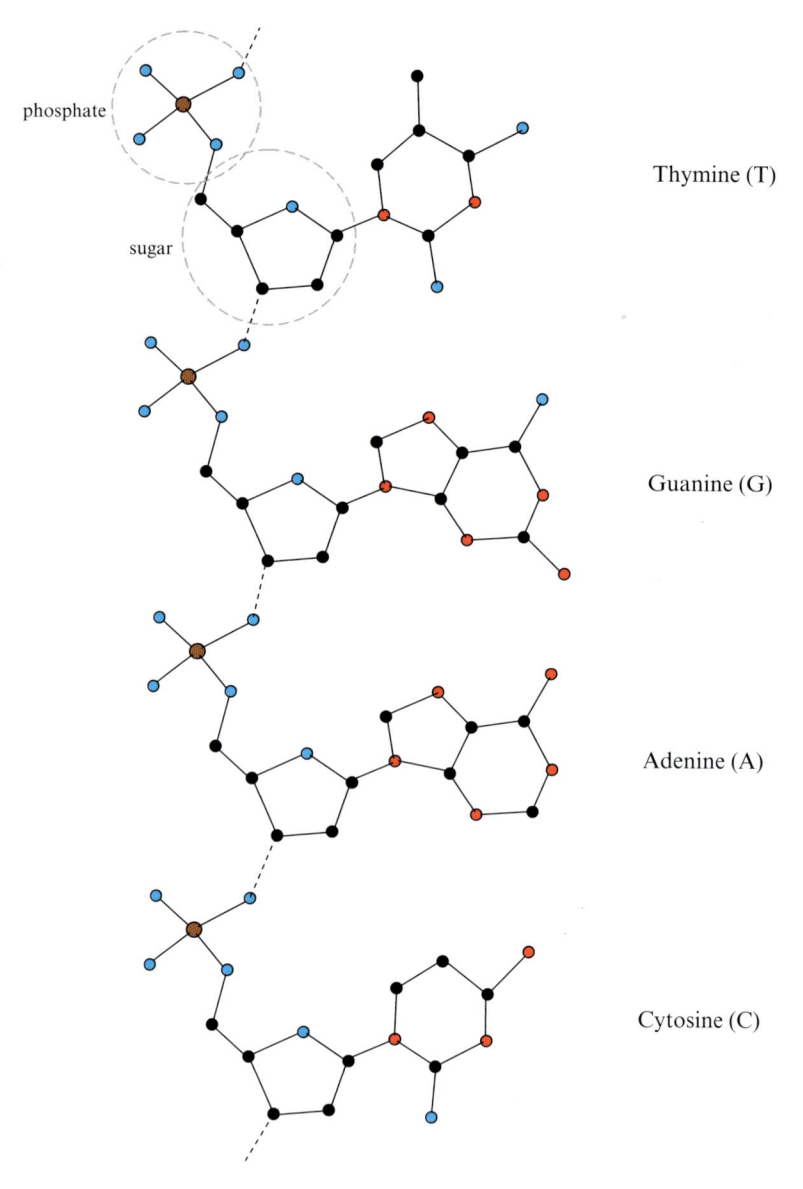

Figure 20.2 The nucleotides that form DNA, each consisting of a phosphate group, sugar and base. The form of linkage of nucleotides is shown by the dashed chemical bond. The atoms present, excluding hydrogen atoms which are not shown, are:

● carbon   ○ oxygen   ● nitrogen   ● phosphorus

largest human chromosome contains 220 million bases. The way in which this chromosome structure influences heredity will be described in §20.5.

Enzymes, which are proteins, read the coded information in the DNA and then form the messenger mRNA referred to in §19.2 in relation to bacterial ribosomes. There is some chemical similarity between mRNA and DNA; the sugar part is different and it contains another pyrimidine, *uracil*, which it uses instead of thymine. Proteins, the complicated molecules that control most of the activities of an organism, are also polymers with the individual units forming the long chains being one or other of 20 *amino acids*. A hypothetical, and impossibly short, protein is illustrated in Figure 20.3.

One unit of the protein chain is indicated within the dashed box. The letter R represents a chemical group, or *residue*, the nature of which determines the particular amino acid at that point of the chain. Figure 20.4 shows the residues for the amino acids glycine, alanine and lycine. At the two ends of the protein the arrangement of atoms is that appropriate to an isolated individual amino acid and they seal off the protein to give it stability.

The protein produced by a particular mRNA is controlled by the sequence of bases. Consecutive triplets of bases in the mRNA, called *codons*, indicate the next amino acid to add to the protein chain and in this way the whole protein is constructed. The human body contains about one hundred thousand different proteins, some being the

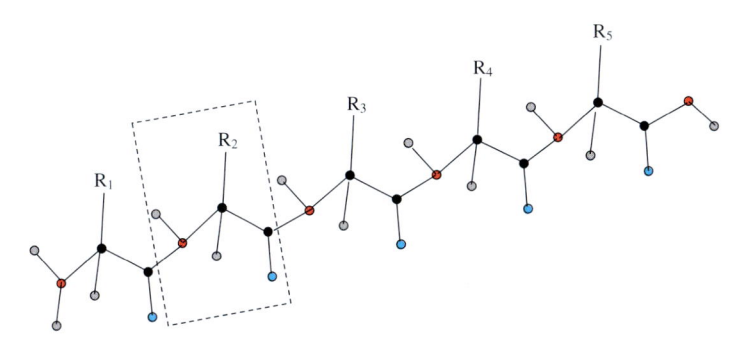

Figure 20.3 A hypothetical protein chain. Atoms are indicated as in Figure 20.2 with o = hydrogen.

glycine   alanine   lycine

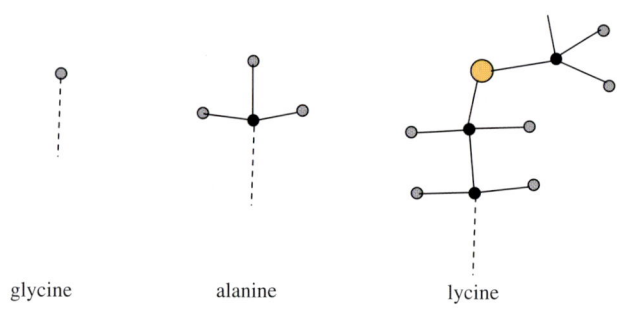

Figure 20.4 Three amino-acid residues. Atoms are indicated as for Figure 20.3 plus ● for sulphur.

physical substance of the body, such as muscle, and others, such as enzymes, which control its activity. Actual proteins can vary in size from having a few tens of amino acids to having many thousands. For example, the protein haemoglobin, the component of blood that is responsible for transmitting oxygen throughout the body, contains four identical protein chains each containing 287 amino acids curled up into a globular form. On the outside of each glob is attached a *haem group*, a small group of atoms containing an iron atom. When blood passes through lung tissue the globs take up a shape that enables oxygen to attach itself to the haem groups. On reaching muscles the globular shape changes again, oxygen required by the muscles is deposited and carbon dioxide, a waste product of muscular activity, attaches itself to the haem groups. When the blood again reaches the lungs the globular shape changes once more, carbon dioxide is released and oxygen is picked up. Then the carbon dioxide is exhaled and the cycle begins again.

The genome controls not only the species of the organism it represents — e.g. an oak tree or an elephant — but also the characteristics of the individual member of that species. The study of this area of science is known as *genomics* — determining the complete, or nearly complete, genomes of many species, varying from viruses (that may or may not be living) to *Homo sapiens*. Comparison of the DNA of different species, different in the sense that they cannot interbreed, shows relationships that indicate the pathways of Darwinian evolutionary processes. Thus there are similarities between the genomes of mice and rats that indicate some common point in their evolutionary pathway between 12 and 24 million years ago. Another striking similarity is that between the genomes of man and the chimpanzee, the two most developed species of the primates. They are identical in 99% of the most important parts of their DNA sequences and the identity is still 96% if one takes the whole genome consisting of 3 billion base units. This comparison gives strong support to the idea of a common ancestry of the chimpanzee and man.

We have given a rather simplified story of the chemical processes that occur in an organism; the details of how the processes occur are complex in the extreme. However, the story is not complete — what is missing is the form in which DNA exists.

20.4 The Double Helix

There are a number of scientific terms that have become so well known that they are almost a part of everyday language. One of these is $E = mc^2$, Einstein's expression for the equivalence of mass and energy, and the other is the term *double helix*. This relates to the form in which DNA exists, not as a single polymer strand but as two strands wound together in a helical form. The elucidation of this structure, involving four individuals — James Watson, Francis Crick, Maurice Wilkins and Rosalind Franklin (Figure 20.5) — was one of the greatest scientific achievements of the 20th century.

In King's College, London University, Wilkins and Franklin had prepared samples of DNA in the form of fibres. While these fibres were not crystalline in a normal sense, the regularities in the arrangement of atoms within them gave rise to an X-ray diffraction pattern showing interesting features (Figure 20.6). This DNA diffraction pattern was a spectacular achievement since nothing like it had been produced previously. While this work was being done in London, at the Cavendish Laboratory in Cambridge Crick and Watson were joining together, with barrel connectors, simulated chemical units, consisting of welded metal rods, to build trial models of DNA based on knowledge of the molecular forms of the constituent nucleotides.

Figure 20.5 Contributors to the 'double helix' discovery from left to right — the Cambridge pair, James Watson and Francis Crick, and the King's College pair, Maurice Wilkins (courtesy Mrs Patricia Wilkins) and Rosalind Franklin.

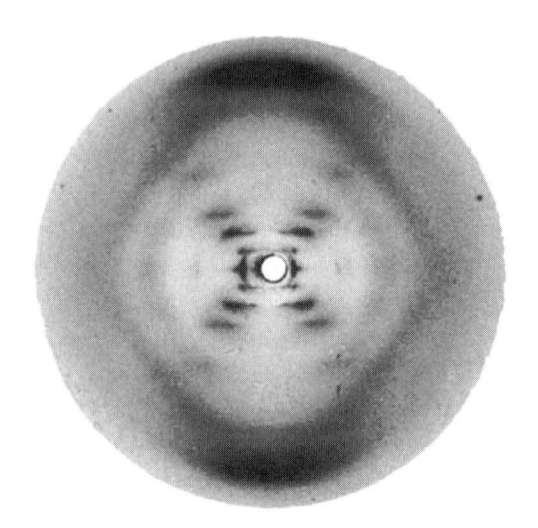

Figure 20.6 Rosalind Franklin's X-ray picture of DNA.

On a visit to King's College, Crick and Watson were shown Rosalind Franklin's X-ray picture by Maurice Wilkins. This gave the Cambridge pair all the information they needed to bring their modelling to a successful conclusion. The X in the middle of the picture indicated that the structure was in the form of a helix and the equi-spaced horizontal streaks that formed the X told them that the bases in DNA were parallel to each other with a particular spacing and also perpendicular to the axis of the helix. There were other facts about DNA that were known to Crick and Watson apart from what was contained in the X-ray picture. When DNA from various sources had been chemically analysed it was found that the amounts of the bases adenine (A) and thymine (T) were always the same, as were the amounts of guanine (G) and cytosine (C). For these pairs one was a purine and one a pyrimidine and Watson, using cardboard models, showed that pairs A + T and G + C could chemically bind together with linkages known as hydrogen bonds in which hydrogen atoms act as a glue holding the bases together. The bonding of A + T and G + C is shown in Figure 20.7. The reason for the equality in amounts of A and T and of G and C was because they were always linked together in pairs in the way shown.

With this information, and a great deal of imagination and ingenuity, Crick and Watson were able to build a model of DNA that was convincing and also explained how it was that DNA could faithfully reproduce itself. The starting point is two parallel chains of nucleotides. The two chains are joined together by hydrogen bonds between bases, with linked base pairs being either A – T or G – C. Now the chains are twisted in such a way as to make the base pairs parallel

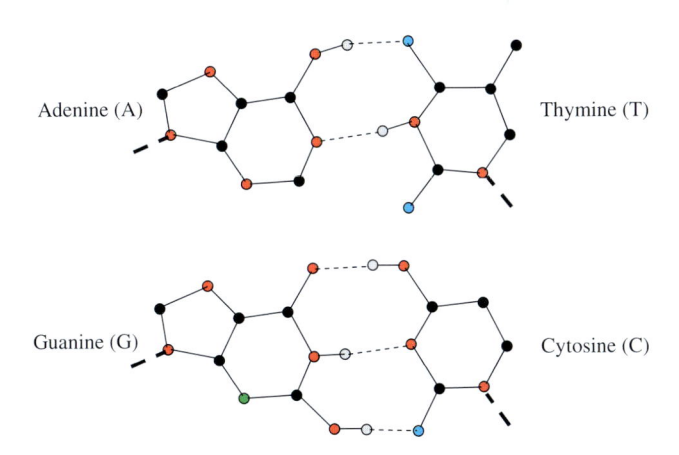

Figure 20.7 The linking of the base pairs A + T and G + C by hydrogen bonds, shown as faint dashed lines. The heavier dashed lines are chemical bonds to the phosphate-sugar chain.

to each other and perpendicular to the axis of the helix. It sounds very simple but it was actually a considerable achievement, bringing together all that was known about DNA at the time together with the experience in model building that Crick and Watson had gained over a considerable period.

In 1953 the structure was published in the journal *Nature* and in 1962 Crick, Watson and Wilkins were awarded the Nobel Prize for Medicine. Sadly, in 1958, Rosalind Franklin died at the early age of 37 so her important contribution could not be considered in the allocation of the prize. However, her contribution is widely recognized and every year the Royal Society makes the Rosalind Franklin Award to a female scientist or engineer for an outstanding contribution in some area of science, engineering or technology. She is also commemorated by the Rosalind Franklin Building, a residence for graduate students, at Newnham College, Cambridge, where she was an undergraduate, and in the name of a prominent building on one of the King's College, London, campuses — the Franklin-Wilkins Building.

A picture of a section of DNA is given in Figure 20.8. The phosphate + sugar backbone gives rise to the helical structure and the sequence of parallel linked base pairs are in the form of an inner spiral staircase that completely define the organism to which they belong. It might be wondered what the significance is of the *double-helix* form of DNA; although many of the products of nature have an intrinsic beauty, they must also be practical and useful. The genetic message is contained in one strand of DNA so why have two? It turns out that the two-strand structure is *critical* if a species is to continue its existence. For an organism faithfully to pass on its genetic structure to the next generation, new DNA must be created from existing DNA *without introducing any errors in the reproduction process*. The way this happens is that a helical section of DNA unravels into its two component strands in the presence of a source of the different nucleotides. As it unravels nucleotides attach themselves to each strand according to the base-pair partnerships. In this way two complementary strands are attached together which, as they form, rewind into a helix. In place of the original DNA helix there are now two helices, each a

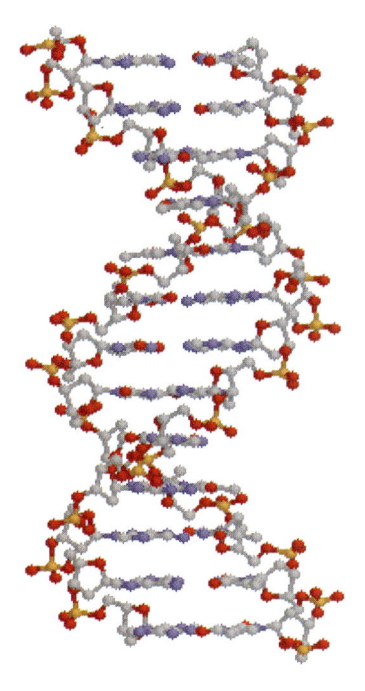

Figure 20.8 Part of a DNA helix.

precise copy of the original one. In principle this copying process can be repeated indefinitely and so the species can survive.

20.5 What Makes Us How We Are?

Figure 20.9 shows the full complement of the 23 pairs of human chromosomes and it is important to consider the significance of the pair structure. We will take one human characteristic, eye colour, and see how the genes contained within the chromosomes influence it.

For the purposes of this discussion we shall make the simplifying assumption that there are just two possible eye colours — brown and blue. The gene that controls eye colour is situated in chromosome pair 19. The chromosomes are in pairs, called *autosomes*, because as you go along the pair you meet genes in the same sequence that both have an influence on the same characteristic. For example, the 159th

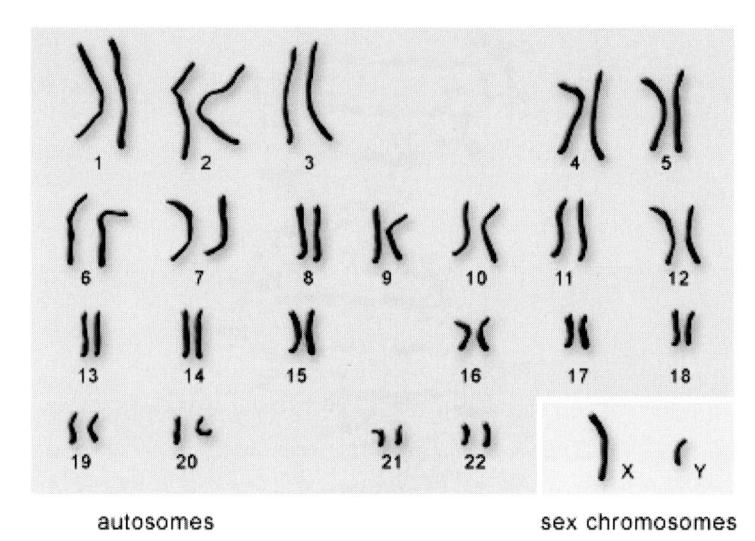

Figure 20.9 The 23 pairs of human chromosomes (US National Library of Medicine).

gene along both the left-hand and the right-hand chromosomes in pair 19 starting from one end both control, let us say, eye colour. If they both are indicating brown for the eye colour then the individual will have brown eyes. Similarly if they both indicate blue for the eye colour then the individual will have blue eyes. However, they may not necessarily be indicating the same thing — so what happens then? In this case the individual will have brown eyes because the brown eye gene is *dominant* and the blue eye gene is *recessive*. The alternative forms of the genes found at the same point of a pair of chromosomes are called *alleles*. This is the basis of inheritance as studied by the Austrian Augustinian priest, Gregor Mendel (1822–1884) who deduced the rules by which inherited characteristics were acquired by breeding generations of peas. In bisexual breeding the chromosomes are inherited randomly from each parent and this gives huge variety in the possible offspring of a single pair of parents. For example, two brown-eyed parents can give rise to a blue-eyed child. For this to happen both the parents must have brown + blue eye-colour gene combinations and the child must have inherited the blue gene from each parent. However, if both parents have blue eyes, and so have blue + blue recessive eye-colour genes, then there is no possibility that their child could have brown eyes.

The story of inheritance is not always as straightforward as described here — there are in fact other eye colours that muddy the simple picture we have given — but the basic mechanism is as described. There is one pair of chromosomes that are different, the sex chromosomes marked as X and Y in Figure 20.9. Females have two X chromosomes and males have X + Y. When children are produced it is the male partner that determines the sex of the offspring since the female can only provide the X variety of chromosome. When Anne Boleyn was beheaded in 1536 at the behest of King Henry VIII on the spurious grounds of adultery, it was really because she had not presented him with a male heir — only a daughter, the future Queen Elizabeth I. Now we know who was really to blame!

Although we have described some of the chemical bases of life — in particular DNA and proteins — there are still many more processes involved in the creation of new life from existing life and these are not completely understood. For human reproduction the cell that begins the new life contains a random selection of the parental genes. This cell divides to produce new cells and this process of multiplying cells, each containing identical DNA, continues as the new individual develops. Then something remarkable happens — the cells begin to differentiate to produce different parts of the new human organism. Some cells become left-arm cells and others right-arm cells and unless something goes wrong the individual will eventually be born a normal human being with the right number of arms, legs, fingers *etc* all in the right positions and with a brain and a nervous system all functioning normally. However, things can occasionally go wrong and children are sometimes born with handicaps. If the chemistry of the mother's body is not normal then the growing foetus may be affected — which is why expectant mothers are encouraged not to smoke or consume alcohol. In the 1960s a new sedative drug, thalidomide, was introduced into the British pharmacopoeia and was prescribed to some expectant mothers to treat stress. Subsequently large numbers of babies were born with stunted limbs; the thalidomide had interfered with the system that controlled the formation of the infant in the womb. While we understand a great deal about the relationship between DNA, genes and the final organism that is produced we still have incomplete understanding of the mechanisms that operate to produce that relationship.

There is a way in which the normal pattern of reproduction can be disturbed. Due to an error in the DNA copying process — which can be due to exposure to radiation, damaging chemicals or even a virus — the DNA passed down to the progeny may be different from the original DNA of the parents. Normally such a *mutation* is harmful to the affected individual and eventually Darwinian selection will eliminate it from the gene pool of the population. However, once in a while such a mutation can confer an advantage to the affected individual and in this case Darwinian selection will enhance the proportion of the mutated gene, perhaps to the point of eliminating the original gene that it replaced. After several substitutions of original genes by mutations that conferred an advantage, what may emerge is a new species with quite different characteristics from that which originally existed.

20.6 The Artificial Manipulation of Genes

There are many practical areas of life where choosing favourable genes over unfavourable ones is a common practice. This activity is not described in terms of genes but is given the name *selective breeding* and is an accepted and respectable activity. Those that breed racehorses pair stallions and mares that share some desirable characteristic, such as stamina, in the hope that their offspring will share that quality and perhaps even have it to an enhanced degree. By selective breeding it is also possible to produce some kinds of agricultural plants that will have advantages — for example, that are drought resistant. There was also an approach, described as *eugenics*, to apply this principle to producing humans with improved desirable qualities. The National Socialist party that ruled in Germany from 1933 to 1945 introduced this as an official policy, coupled with the idea that there was a Nordic master race that should be especially promoted. Despite the fact that eugenics has had many famous supporters in the past, as diverse as Woodrow Wilson, George Bernard Shaw and Winston Churchill, it is now looked down on as an unworthy activity for humankind.

Biotechnology has now advanced to the point where individual genes, conferring particular characteristics on an organism, can be identified and, if desired, replaced. This ability has raised many possible applications and also many ethical issues. Perhaps the least controversial application is *gene therapy*, the cure of disease by replacing a faulty gene in the patient's body by a normal one. The common way of doing this is to insert DNA containing the normal gene into a virus that has been genetically modified to contain human DNA. The malfunctioning component of the body, say the liver, is infected with this virus that then replaces the defective gene by a normal one.

More controversially, there is the selection of an embryo by *in vitro fertilization*, fertilization of an ovum by sperm outside the womb, to obtain a child that will have suitable genetic material to transplant to an older sibling that has a disease due to a faulty gene. To some people this kind of selection smacks of eugenics; others argue that if the parents' intention is to have another child anyway then it is sensible to have one that could improve the life of the previous child.

A final area of controversy is the production of genetically modified (GM) food. Plant breeding by natural means has a long history but in the 1990s various food plants were produced that had genes replaced artificially, sometimes from animal sources, to confer beneficial properties to them. Plants such as maize and soya were produced that gave higher yields, were resistant to drought or would produce their own insecticide so eliminating or reducing the need to spray crops. There was no objection as such to the actual improvement of the plants — indeed it would have been welcome if the new varieties had been produced by natural means — but concerns were expressed that the new plants might be harmful to human health. So far there is no evidence of harm ensuing from eating GM foods but many countries have stringent regulations that prevent or heavily control their production and distribution.

The ability of people to manipulate the gene structure of plants and animals, including the human animal, has already thrown up many ethical issues and will continue to do so in the future.

Chapter 21

Materials From Ancient Times

21.1 The Earliest Use of Natural Materials — The Stone Age

It sometimes seems that humanity, or at least a less thoughtful part of it, regards humankind as an observer and interpreter of the natural world but not as a component of that world. That is a comparatively recent state of mind, where the term 'recent' is in relation to the lifetime of man as a species. We can live in a warm environment at cold times of the year and in a cool and comfortable environment when the weather is hot and humid. We can perform great feats of engineering, changing the landscape for our own convenience. The depths of the ocean and the heights of space are ours to be explored if that is what we wish to do. Other species of the living natural world must put up with the conditions to which nature subjects them and if they cannot tolerate those conditions then the principles of Darwinism will operate and they will have either to evolve to meet the new conditions or will become extinct.

To our forebears, twenty or thirty thousand years ago, the place of humankind in the world would have seemed very different. The human animal had to survive in the presence of many other animals that possessed enormous physical advantages. Man was no match in strength for a mammoth or a sabre-toothed tiger and no match in speed for most of the animal kingdom with which it had to compete. Only in one respect did man have an advantage and that was in intelligence, made possible by his highly developed brain. By the use of his superior intelligence man was able to exploit the natural materials available to him both to improve his performance as a predator and to increase his chances of survival as potential prey.

The use of natural resources is not confined to man since many animals also show an ability to use what is available. Some seabirds will drop shellfish from a great height onto rocks to crack them open and seals also use stones to crack shells. Dolphins — well known for their intelligence — hunt for fish off the coast of Western Australia by using their snouts to stir up the seabed where many smaller fish take refuge. However, in those waters they are in danger of being injured by spinefish, which are bottom-dwellers and have poisonous spines protecting their bodies. To shield themselves against this hazard the dolphins pick up sponges to cover the front of their faces and they select sponges of a conical shape that are retained on their snouts more securely. Chimpanzees, man's closest relatives, are also known for their ability to exploit natural objects to achieve their ends. Experiments have been done with chimpanzees in captivity in which bananas have been placed out of reach of the caged animals. If a chimpanzee has a stick available then it will use it to slide the bananas closer in, to within reach. Again, if bananas are placed at an inaccessible height then chimpanzees will pile up boxes to climb on to reach the fruit. In the wild, chimpanzees have another interesting way of turning a natural object into a tool. They will prod into a termite nest with a stick, or sometimes a stiff blade of grass. When the prod is withdrawn there are many termites on it — a tasty snack for the chimpanzee. Sometimes the chimpanzee will chew the end of the stick, which increases the yield of termites. Some birds — such as woodpecker finches and green jays — also use sticks to winkle grubs and insects out of crevices.

The first exploitation of natural objects by man would possibly have been the use of stones and rocks as projectiles and to crack the shells of shellfish, and of heavy branches as clubs. Broken stones would have provided sharp edges that could be used as crude knives and over time the craft of fashioning tools and weapons such as knives and axes by using stones to break each other would have been developed. Fixing a sharp stone to a long pole, derived from a straight branch of a tree, would give extra reach and enable the weapon to be used out of range of the fangs, claws or horns of the beast being attacked or doing the attacking. The ability to learn and to adapt, which is man's heritage, would gradually have increased the sophistication of the tools and weapons being made, a trend that continues to the present day. However, one of man's early, and possibly greatest, achievements is his ability to control and use fire, a skill acquired more than 200 000 years ago. Natural fires, caused by lightening or volcanoes, occur from time to time and are instinctively feared by most animals. Mammoth bones found at the bottom of cliffs have suggested that a hunting technique used by early men was to panic the animals by means of waving burning brands, so driving the animals to their deaths over the edge of the cliffs. By the use of tools and fire, early man was gradually obtaining mastery over the rest of the animal kingdom, although the world was still a very dangerous place.

The earliest materials used by man were those directly offered by nature — wood, stone, clay, animal products and reeds, for example. A Stone Age man could hunt and gather plant food, he could mine flints with deer-antler pickaxes to make tools and weapons, he could make a shelter using some combination of stone, wood, animal skins and clay, he could make pottery from clay and clothing either from animal skins or even by primitive weaving of animal or vegetable fibres. To this catalogue of resources must be added the control and use of fire. This was the launching pad from which humankind advanced by small steps up to the present time, gradually extending its knowledge of how to exploit the material — the matter — that is contained in the Earth on which it lives.

21.2 Some Early Manufactured Materials — The Bronze Age

While archaeology can tell us at what times in history, or pre-history, various technologies were present it is difficult to know exactly how and when they began. There are some very plausible theories for *how* they may have begun, mostly based on chance occurrences that

216   *Materials, Matter and Particles*

produced a useful material and also indicated to an astute observer how the process could be reproduced at will. A notable example of such a chance occurrence is probably to be found in the discovery of, and subsequent production of, glass.

Glass is best described as a 'frozen liquid'. In crystalline materials, like the sodium chloride illustrated in Figure 18.14, atoms are fixed and arranged in a regular pattern, like three-dimensional wallpaper, over distances that are very large compared with the repeat distances of the pattern. This contrasts with a liquid in which atoms are in constant motion. At any instant there is some approximate local order in the atomic arrangement but this order does not persist over large distances. Figure 21.1 gives an impression, in two dimensions, of the difference between a crystal and a liquid arrangement of atoms. In a glass the atoms are more-or-less fixed, as in a crystal, but have the lack of order in their positions as for a liquid. The term 'more-or-less fixed' to describe the positions of the atoms in glass was used for a reason. In reality glass is not a solid but a liquid — one that flows so slowly that for all practical purposes, over periods of time much greater than a human lifetime, it behaves like a solid. The liquid properties of glass can be confirmed by examining glass window panes in very old buildings. Under the influence of gravity the glass flows downwards and the glass at the bottom of the pane is found to

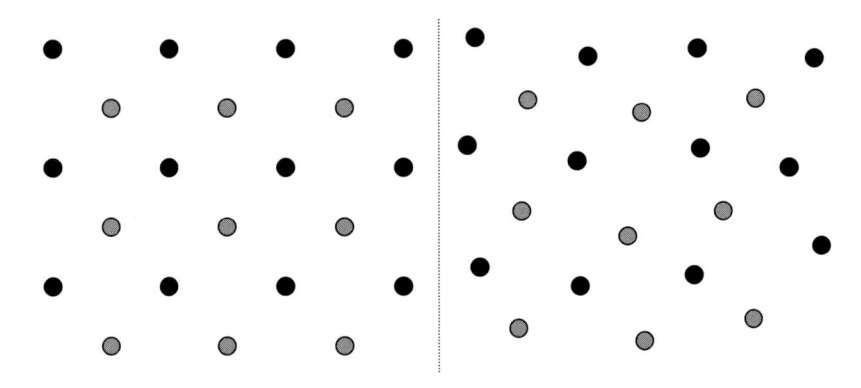

Figure 21.1 Representations of a crystal structure (left) and a liquid structure (right).

be thicker than at the top. This is a manifestation of the flow of a liquid with a very large *viscosity*. Water, which flows easily, has a low viscosity, treacle, which flows less well, has a considerably larger viscosity. Tar, the material sometimes used in road building, flows very reluctantly and hence has a very high viscosity — although in unusually hot weather the tar becomes more fluid and the road surface may then become distorted. Glass has an extremely high viscosity — so great that its flow properties cannot be discerned over a short time span.

The main ingredient for making glass is sand that, chemically, in its very pale form, is almost pure silicon dioxide, SiO_2, a material known as *silica*. Another form in which silica occurs is in the mineral quartz that occurs in the form of large crystals. In fact, glass can be made of silica alone by melting sand or quartz, which melt at a temperature of more than 2 300°C, in an electric furnace and then allowing the melt to cool. This so-called *silica glass* has very specialized uses but is too expensive for everyday use in making common objects like bottles. However, mixing sand with sodium carbonate, Na_2CO_3, produces a glass, *soda glass*, with a melting point of 1 500°C, a temperature that is attainable with a bellows-assisted normal combustion furnace. A disadvantage of normal soda glass is that it is readily leached by water and so degrades. To make it more durable, without greatly affecting the melting point, small amounts of oxides such as calcium oxide (lime), CaO, magnesium oxide, MgO, or aluminium oxide, Al_2O_3, are added to give what is known as *soda-lime glass*.

Some glass occurs naturally, such as the mineral *obsidian* that is produced by the melting of some kinds of rock during volcanic activity with subsequent rapid cooling that prevents the formation of crystals. Figure 21.2 shows an example of obsidian from Oregon in the USA. It is capable of being ground to give a very keen cutting edge and so it was much prized in some early societies for the making of knives and weapons of various kinds. Indeed, even in the present age some surgical scalpels are fashioned from obsidian.

One can only speculate about the origin of man-made glass. A fire lit in a sandy desert in the presence of soda, which can be derived

Figure 21.2 A lump of obsidian from Oregon.

from the ash of burnt plants, plus a wind that is funnelled through the fire to give a high temperature, could lead to globules of opaque glass being formed. The making of glass probably originated in Mesopotamia, modern Iraq, about 2500 BCE, and was first used for making beads and various decorative objects. About 1 000 years later the first glass vessels were produced by covering a clay core with a layer of glass, shaping and smoothing the glass layer and then, when the glass had cooled, removing the brittle clay interior. Later, in Syria or thereabouts, the art of glassblowing was invented. This involved placing a blob of molten glass on the end of a tube and then, by blowing through the tube, creating a glass bubble that could be fashioned into vessels of any required shape.

The science of glassmaking has progressed through the ages so that, from its origin as an expensive luxury product, glass has now become a common and cheap material. Bottles for various grocery products and for wine are produced by the hundreds of millions and are then discarded, although in this environmentally conscious age they are recycled and made into other containers or turned into home-insulation glass-fibre sheets. Specialist glass, some of it quite expensive, is also made in some quantity. In the 1950s the UK glass manufacturers, Pilkington Brothers, invented the process of making *float glass* that involved depositing molten glass onto a surface of molten tin. Through the action of gravity the glass is flat and uniform and the process can be designed to make large continuous sheets by having molten glass deposited at a uniform rate at one end of a molten tin bath and the solid glass sheet being drawn out from the other end. This process can be used with soda-lime glass for ordinary windows or with special toughened glasses where there are security or safety requirements.

Predating the discovery of glass was the discovery and use of metals. The ones that were first used were those that occur in their basic metallic form in nature — copper and iron. Copper is found in its raw native state in basaltic rocks derived from volcanic activity. This copper would have been produced by chemical processes, occurring in the high temperature of a volcano that extracted copper from one of its ores. An example of a lump of copper produced in this way is shown in Figure 21.3.

The use of this readily available material dates back at least 10 000 years. The earliest copper artefact, from Mesopotamia, is in the form of a copper pendant and is dated about 8700 BCE. Copper is a rather soft metal and hence has a limited use for weapons or tools. However, copper chisels have been found in ancient sites and were used in working stone for the building of the Egyptian pyramids. In 1991 the mummy of a man, preserved in ice was found in the mountain region that separates Italy and Austria. Ötzi the Iceman, as he has come to be called, has been dated to between 3300 and 3200 BCE. Amongst the possessions of the 'Iceman' were a copper axe with a yew handle

Figure 21.3 A naturally-occurring lump of raw copper.

and a flint knife with an ash handle plus a bow and arrows with flint arrowheads; his clothes were made of fur and leather. This find indicates that more than 5 000 years ago the use of copper had intruded into the Stone Age and that a widening range of materials was being used by the people of the time. The earliest use of copper seems to have been in the Balkans, Turkey and the Middle East but later uses have been recorded in Northern and Western Europe, China, the Andes region of South America (Chile is an important modern source of copper), Central America and West Africa.

The most detailed knowledge of the early use of copper comes from ancient Egypt. The Egyptians were making copper beads, ornaments and tools during the fourth millennium BCE. Some of the copper they used would have been native copper but it has been suggested that they also extracted copper from the mineral *malachite*, the green carbonate of copper that is abundant in Egypt and was used as a cosmetic in the form of eye paint. It might also have been used to decorate glazed pottery and in a charcoal-fired kiln some malachite could have been reduced to metallic copper, so leading to the discovery of the smelting process for extracting the metal.

It was usual to make copper artefacts by pouring molten copper into moulds — but this process produced a metal containing bubbles that would therefore be somewhat brittle. The bubbles could be removed by hammering the material and this had the additional advantage that the copper *work-hardened*, thus improving its usefulness for utilization in weapons or tools.

In the region of modern Turkey, and in the North Caucasus, copper minerals are sometimes found in intimate contact with arsenical minerals and smelting this combined material produced a copper-arsenic alloy, *arsenical bronze*, which is vastly superior to pure copper in its physical properties. Later it was found that adding tin to copper gave another form of bronze with similar properties and this became the material of choice in the ancient world for the manufacture of tools and weapons. Whereas copper tools and weapons lost their keen edges quite quickly, so that in many ways stone implements were preferable, the bronze tools stayed sharp for prolonged periods of use. In most early civilizations there was a considerable period of time when bronze was the dominant metal being used. The *Bronze Age* started and ended at different times in different parts of the world — dictated to a great extent by the availability of the basic ingredients for making the alloy. The earliest region within which bronze was manufactured and used on a large scale was in the Indus Valley, where the Bronze Age started in about 3300 BCE and lasted until about 1500 BCE, at which time iron took over as the dominant metal. In the European region the Bronze Age began with the Minoan civilization, based on Crete, in about 3000 BCE. The main source of copper in the region was Cyprus — from which copper obtained its name, *cuprum* in Latin, giving rise to the chemical symbol Cu. Tin, and later charcoal when Cypriot forests were being exhausted, was imported into Cyprus where there was a thriving bronze industry, the products of which were exported into much of Europe and the Caucasus region. Some of the tin was obtained from Cornish mines, where tin was mined up to the end of the 20th century. In 1900 BCE an indigenous British bronze industry was set up using local sources of copper as well as tin (Figure 21.4).

Figure 21.4 Bronze castings of Celtic origin (Somerset County Museum, Taunton).

Almost contemporaneous with the European Bronze Age, the use of bronze was being developed in Egypt. Chemical analysis of a bronze statue dating from the reign of the Pharaoh, Pepi, around 2300 BCE found the presence of Cornish tin, indicating the extensive trading networks that must have been operating at that time. The Bronze Age reached China about 2000 BCE and also occurred in the Andes region of South America about 900 BCE.

The discovery and use of bronze represented a major advance in the exploitation of the Earth's resources by humankind. However, in about 1500 BCE its pre-eminence was terminated by the production of another, even more useful, material — iron.

21.3 The Iron Age

Iron of terrestrial origin does not occur in its native state but always as ores of one kind or another, usually as oxides such as *magnetite*, Fe_3O_4, or *haematite*, Fe_2O_3. Despite this, the use of iron probably goes back many thousands of years because sources of native iron *are* available on Earth — that contained in iron meteorites (Figure 21.5).

The largest iron meteorite known is that discovered in Greenland in 1894 by the American Arctic explorer Robert Peary, who was following up an 1818 report by an earlier Scottish explorer, Sir John

Figure 21.5 An iron meteorite (NASA).

Ross, that Inuit people in Greenland were using knives made of meteoritic iron. The meteorite had broken into seven pieces, three of which, named *Ahnighito* (31 tonnes), *Woman* (3 tonnes) and *Dog* (400 kg), were found by Peary — the others at some distance were discovered later. These meteorites had been used for centuries by the local Inuit people as a source of iron for the making of harpoons and other tools. The Inuit would have not been able to forge the iron since they could not have produced the necessary high temperatures but meteoritic iron, which contains from 7½% to 10% of nickel, is fairly soft and can be worked by cold hammering.

Evidence of the use of, or knowledge of, meteoritic iron is known from several ancient sources. A meteoritic iron dagger was found in the tomb of the boy Pharaoh, Tutankhamun (1341–1323 BCE). An iron meteorite falling at Kaalijärv in Estonia about 2500 years ago broke up and formed a series of small lakes. People in Estonia made iron tools and weapons and it is a reasonable assumption that their source of iron was meteoritic.

The big breakthrough in the use of iron came about when the process of smelting iron ores was discovered. Smelting, as a process of extracting metal from metallic ores, is similar for many metals. If the ore is an oxide of the metal, such as magnetite or haematite, then heating it in the presence of carbon, which in ancient times would generally have been charcoal, removes the oxygen in the form of carbon monoxide, CO, or carbon dioxide, CO_2, leaving behind the metal. Usually the metallic ore is mixed with rocky impurities and these do not melt at the temperature of molten iron and so remain in solid form and float on the surface, from which they can be skimmed off. By adding limestone ($CaCO_3$) to the mixture the rock breaks down to calcium silicate ($CaSiO_3$) that melts and is left behind as a *slag* floating on the surface of the molten metal. Removing this slag then leaves behind the metal, which will be purer than that obtained by removing solid rock.

The problem for ancient peoples with the smelting of iron is that it requires a much higher temperature than those needed for the other metals produced hitherto, such as copper, tin, lead, mercury and silver. The discovery of smelting as a process probably came about by accident. Metallic ores are in the form of rocks and if a campfire had been built over copper or tin ores, or a stove was built of such material, then metal may have been produced since the temperatures required for smelting these ores are readily achievable in a camp fire or stove in favourable circumstances, such as in a high wind, and the fuel itself would have provided the source of carbon. Once the smelting process had been discovered then its development would have continued; ancient man was no less capable of inventiveness than we are today. It would soon have been found that higher temperatures gave a better result and the experience of seeing enhanced combustion in a wind would have led to some kind of bellows system for producing an artificial wind.

The first production of iron could have come about as a by-product of copper smelting with some iron ore present. The iron so produced would have been of a spongy nature and of very poor quality but from that starting point, by small steps, the production of useful iron would have been an inevitable development. Iron technology grew up in the period known as the *Iron Age*, generally taken as beginning about 1200 BCE, but its beginning varied greatly in different parts of the world. It began in coexistence with the general use of bronze, probably as an exotic and expensive material since it was comparatively difficult to produce. However, once the technology of producing iron was well established it quickly displaced bronze as the material of choice because it had certain technical advantages, i.e. it could be sharpened by grinding whereas a bronze tool or weapon needed re-forging, and also because iron ores were more plentiful than ores of copper and tin. Another factor is that by about 1800 BCE there seems to have been a shortage of copper and tin and the cache of bronze castings seen in Figure 21.4 was probably to be used for recycling into other bronze objects.

The assessment of when iron technology became established in various parts of the world depends on the discovery of artefacts made of iron and may require reassessment when new discoveries are made. Present evidence suggests that India, in the region of the province of Uttar Pradesh, may have been the place where iron technology really began. Some iron implements discovered in this region date from about 1800 BCE and iron smelting was carried out on a large scale by 1300 BCE. The original iron being produced, in India as elsewhere, would have been *cast iron*, which contains about 4% of carbon and is a rather hard and brittle material. If all the carbon is removed then the pure iron that remains is known as *wrought iron*, which is malleable but not very strong. Intermediate proportions of carbon content give *steel*; the general rule is that the hardness and brittleness increases with the amount of carbon. *Mild steel* contains about 0.25% of carbon and is a good general-purpose material combining reasonable hardness and some malleability. Increasing the amount of carbon to 1.5% gives *high-carbon steel* that is much harder than mild steel and less malleable, making it more suitable for some applications. Between 2 000 and 2 500 years ago the Iron Masters of southern India were producing steel by heating together wrought iron and carbon under well-controlled conditions. From India the knowledge of controlled steel making gradually permeated into other parts of the world.

There is a common misconception that the rise of technology was confined to the civilizations of Asia, the communities that ringed the Mediterranean Sea, Central and Northern Europe and, to a lesser extent, Central and South America. However, when it comes to the smelting of iron one of the earliest examples of an iron-age culture comes from eastern Niger in sub-Saharan Africa. The production of iron began there in about 1500 BCE and was taken from there by Bantu-speaking people, mainly southward towards modern South Africa. The Zulu nation, one part of this migration, by virtue of its superior iron tools and weapons became the dominant power in southern Africa and later established a large empire there under King Chaka (Figure 21.6).

Another region of early iron smelting, about 1300 BCE, was somewhere in the region bounded by the Caucasus and the northern part of modern Turkey. From here it spread into other areas of Asia, including the Middle East and Egypt and into Europe via Greece. The spread of the technology took some time, reaching both central Europe and Britain in about 800 BCE. For a civilization that was so early and so prolific in its inventions, the Chinese came somewhat late

Figure 21.6 King Chaka (1787–1829) with a iron-bladed spear and large cowhide shield.

into the Iron Age; it is believed that they imported the technology from India in about 700 BCE. From China the art of iron smelting then spread into Korea and Japan.

Although meteoritic iron was used in the Americas, the accepted wisdom is that the smelting of iron was unknown until the arrival of Europeans. The first iron smelting in North America was probably by the Vikings who occupied settlements along the north-Atlantic coast of North America in about 1000 AD. Similarly, the Spanish Conquistadors introduced iron smelting into central and South America in the 16th century — although the smelting of other metals and the production of bronze objects was carried out there long before they came.

In a real sense, because modern society is still highly dependent on iron and steel, the Iron Age can be considered still to continue. Ships, cars, bridges, power pylons, railway tracks and even some buildings depend on steel for their construction. Steel alloys containing small amounts of other metals such as nickel, chromium, molybdenum and tungsten have characteristics that make them suitable for special applications. Thus, alloys containing chromium and nickel give stainless steel, present in many homes as daily-use cutlery. Molybdenum steel is extremely hard and is used for rifle barrels and tungsten steel, with a combination of toughness and heat resistance, is used for the outlets of jet engines. When iron was first produced at the beginning of the Iron Age it was more valuable than gold. While that is not true today what *is* true is that life today could easily survive without gold but could certainly not easily survive without iron, in one or other of its many forms.

21.4 Cement and Concrete

A characteristic of civilized societies of the past and present is the great buildings and structures that they produce. The Great Wall of China, the temples and pyramids of ancient Egypt and the ziggurats built by the ancient Babylonians and Assyrians are a few examples from many that are found in various parts of the world. The earliest building materials were stone and mud bricks but since the fit between the stone blocks and individual bricks was not perfect they were usually separated by some material that served the purpose both of filling in the gaps and binding the components together. A notable exception to this practice is provided by the Inca civilization that flourished in the western parts of South America from the 13th century until its destruction by Spanish invaders in the 16th century. The Inca builders developed the technique of so shaping neighbouring stone blocks in their buildings that it is impossible to even slide a piece of paper between them. How they did this is still not understood.

The ancient Egyptians used a mixture of mud and straw to bind their sun-dried bricks together and they used mortars, or plasters, of both *lime* and *gypsum* in constructing the pyramids. The basis of a lime plaster is lime, calcium oxide, CaO, produced by heating limestone, which is calcium carbonate, $CaCO_3$. Mixing sand, lime and water produces a paste, lime mortar, which sets hard when the lime reforms into limestone. When gypsum, which is hydrated calcium sulphate, $CaSO_4 \cdot 2H_2O$, is heated it loses three-quarters of its water content. This material, mixed with sand and water, reforms the gypsum and also sets hard. The Chinese used mortar in the construction of their Great Wall, as did the ancient Greeks in building their huge stone edifices.

A big advance in building materials was the invention of concrete. This consists of a mixture of cement powder and an aggregate of solid material such as stones, broken bricks, ceramic fragments and the like. When mixed with water the concrete sets into a hard rock-like material but with the advantage over rock that it can easily be moulded into complicated shapes. There is some evidence that a form of concrete was used in the construction of some parts of Egyptian pyramids and, if this were so, then it probably indicates the first use of this versatile material. However, the early masters in the use of concrete were undoubtedly the Romans. Their big advance in concrete technology was to create a concrete that would set under water by using a cement component consisting of a mixture of lime and volcanic sand. A cement mixture with similar properties, known as *Portland cement*, is widely used today in construction work all over the world.

With their mastery over concrete construction the Romans built great buildings, port facilities and aqueducts (Figure 21.7) that have lasted in good condition to the present day. A triumph in the use of concrete was the construction of the domed roof of the Parthenon in Rome (Figure 21.8) with a diameter over 43 metres. As the roof sections proceed from the base towards the apex so the aggregate in the

Figure 21.7   A Roman aqueduct at Segovia in Spain.

Figure 21.8 A painting of the Parthenon showing its domed roof.

concrete is changed to give a decreasing density of concrete as one moves upwards. At the bottom the aggregate is a dense rock giving the concrete a density of 2 200 kilograms per cubic metre while at the top the aggregate is a pumice-like material so that the density is 1 350 kilograms per cubic metre. The denser and stronger material is at the bottom where the greatest weight must be supported and that weight to be supported at the bottom is reduced by using lower-density concrete further up the dome.

The modern use of concrete has been greatly extended by the use of reinforcement — the moulding of the concrete around steel rods that add structural strength to the material. Concrete is a very strong material under compression but much less strong when subjected to stretching forces. In some circumstances concrete may be under extensional stress, for example due to bending, and in such cases the use of steel reinforcing rods enables concrete to be used where otherwise it would be unsafe.

Concrete remains a most important material in the present age and more of it is used than of any other manufactured material. The total produced each year is nearly equivalent to one tonne per person of the world's population. Without it the world as we know it today would not be possible.

21.5 Clothing Materials

For primitive man the essential requirements of life were food, clothing and shelter. In equable climates the need for clothing and shelter was less intense but in areas where extreme cold could be experienced their need was just as urgent as that for food. Many of the animals that were hunted for food had skins, with or without fur, which could be turned into clothing. Indeed, we do not have to go back to prehistoric man to find examples of clothing made from such materials. Although many Inuit people now live in wooden houses and have adopted a settled lifestyle, traditional Inuit who live by hunting animals use the fur of caribou, polar bears and seals to protect themselves against the extreme cold of their environment (Figure 21.9).

Figure 21.9 An Inuit family wearing traditional fur clothes.

Traditionally the skins were scraped and cured and then sown into a garment using whalebone needles with animal sinews as thread.

Despite the simple lifestyle of the Inuit to whom we have referred, they still belong to the modern world. Early *Homo sapiens* and *Neanderthal man* would also have used the skins of animals for clothing but probably in a less sophisticated way — we have no real way of knowing. However, we can get an insight into the clothing of early man from 5300 years ago from that found with Ötzi the Iceman. Most of his garments — a coat, loincloth and leggings — were made of goatskin and constructed of strips of the leather bound with sinews. He had a bearskin hat with leather straps for securing it firmly onto his head. His shoes were quite sophisticated with bearskin soles and deerskin uppers and there was grass inside the shoes acting like socks to keep his feet warm. He also had a cloak made of woven grass.

A big advance in clothing technology was made when the art of weaving yarn to make cloth was developed. The first step is to make the yarn, which is done by twisting short fibres of some material together to make a long continuous thread. This can be carried out in various ways: Figure 21.10, a painting, by the French artist William-Adolphe Bouguereau (1825–1905), illustrates an early method. The forming yarn is attached to a heavy object, the spinning motion of which causes the twisting action by which new yarn is formed. The tufts of fibre that will be added to the growing yarn are stored on the spindle held in the woman's left hand.

The next stage in the production of cloth is to weave the yarn, giving an interlaced set of threads at right angles to each other as shown in Figure 21.11. It is not certain how and when the arts of spinning and weaving were first discovered and used but an intriguing impression in burnt clay found at a Palaeolithic site in the Czech Republic shows the probable existence of a woven material some 25 000 years ago.

The first natural fibres that were used in making fabrics were wool, flax, silk and cotton. The use of wool probably began with the use of sheepskin from wild sheep, killed for their meat. When sheep became domesticated, and kept for their milk as well as their meat, it would soon

Figure 21.10 The Spinner by William-Adolphe Bouguereau.

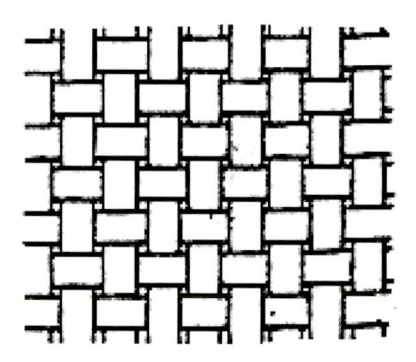

Figure 21.11 A simple woven material.

have become apparent that it was more efficient to take the wool repeatedly by shearing rather than to kill the sheep for a single sheepskin. The earliest woollen garment that has been found was recovered from a Danish bog and dates from about 1500 BCE. However, it is generally believed that the use of woven woollen garments, albeit of a crude kind, probably began before 10 000 BCE although this cannot be directly verified. Wool was certainly an important commodity to the ancient Persians, Greeks and Romans and it is mentioned in Deuteronomy in the Old Testament where as part of Jewish law it was laid down that clothing should not contain a mixture of woollen and linen fibres.

Linen is a cloth woven from the spun yarn from fibres derived from the flax plant. The earliest records that describe the production of linen come from Egyptian accounts in ancient hieroglyphs. Some of the outstanding examples of old linen come from that period; when the tomb of the Pharaoh Ramses II was discovered, the linen bindings of the mummy, more than 3 000 years old, were in perfect condition. Linen found in other ancient Egyptian tombs has been found to be in similarly good condition. The Phoenicians traded linen over the Mediterranean region and beyond. In modern times Ireland is noted for the quality of its linen products and it is believed that the Phoenicians introduced the flax plant and the know-how for producing linen into that country in the pre-Christian era.

The most exotic of all the natural fibres is silk and it is certainly the most expensive. The origin of silk production is uncertain but Chinese legend dates it to about 5 000 years ago. The silkworm — for such we name it now — was a pest in ancient China because of its practice of consuming mulberry leaves. The legend runs that about 5 000 years ago the Empress Si Ling Chi was drinking tea under a mulberry tree in her garden when a silkworm cocoon fell into her cup. When the cocoon fell into the hot liquid it began to unravel and turned into a long strong thread. This is how the process of producing silk is carried out. The cocoons are immersed in hot water and the fine filament that forms it, between 600 and 900 metres long, is loosened and is wound onto a spool. Between five and eight of these filaments are spun together to form the silk thread from which cloth is woven.

The production technique for producing high-quality silk was quite complicated and was a closely guarded secret since it had great commercial value. The silkworm eggs are initially kept at a temperature of 18°C and over several days slowly increased to 25°C at which point the silkworms hatch. They then voraciously feed on a diet of chopped mulberry leaves until they have increased their body mass by a factor of 10 000 at which stage they then start producing the cocoons that are the source of silk. Silk, and other commodities such as gold and ivory, were traded all over the inhabited world by trade routes linking China to South-East Asia, India, Persia and the Middle East, North Africa and, via merchants in the Eastern Mediterranean, as far as Central and Northern Europe. The most famous of these trade routes was the Silk Road that actually consisted of several different routes connecting China to the Eastern Mediterranean, some passing as far south as Northern India while others passed through the countries of Central Asia.

Eventually the secret of silk production leaked out and the Chinese monopoly was broken. In 550 AD two Nestorian monks smuggled silkworms out of China concealed in hollow bamboo staves and offered them to the Byzantine Emperor Justinian. The secret of silk making reached other parts of Europe, and Italy in particular became a major centre of silk production (Figure 21.12).

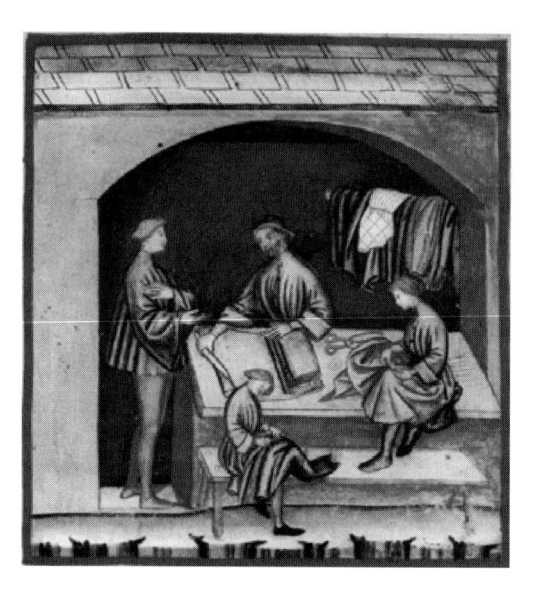

Figure 21.12 The tailoring of silk garments in medieval Italy.

The cotton plant grows in the wild and the appearance of the fluffy bundles of fibres surrounding the seed of the plant, looking just like tufts of wool, must have readily suggested that they could be spun into yarn and used for the production of cloth. The original main areas in which cotton was cultivated were in parts of Asia, in particular the Indus region, and in Central America. The earliest cotton fabrics have been found in archaeological sites in the Indus valley dating back to 3200 BCE. From India cotton textiles spread by trading and reached Rome by about 350 BCE. The arrival in India of European traders, first from Portugal and later from France and England, and the Moorish conquest of Spain were instrumental in the introduction of raw cotton into Europe and the establishment of a cotton industry. Granada in Spain and Venice and Milan in Italy became centres of cotton manufacture but the cotton manufacturing capital of the world was Manchester in England, which began cotton spinning in 1641.

The development of cotton production in the southern states of the USA, plus the labour needs of sugar plantations in the West Indies, were undoubtedly the big drivers of the development of slavery in those areas. It is to the credit of the cotton workers of Lancashire that they refused to support the South in the American civil war even though the consequent shortage of cotton brought them great hardship.

Chapter 22

Modern Materials

22.1 Materials Old and New

We have seen how from the very earliest time man has exploited a wide range of materials to improve his everyday life. Some of these materials, such as wood, stone and animal and vegetable fibres, were in their natural form while others, such as metals extracted from various ores or glass and concrete, which are composite materials with several components, required some sort of manufacturing process. These materials, the use of which originated in antiquity, are still important in the modern world. Homes today have components built of stone and wood, windows and vessels made of glass, metallic objects, many made of the ancient metals, and concrete in the structure of the buildings themselves. The clothes that people now wear still contain the fibres used for the clothes of long ago, although modern manufacturing processes may give the fabrics a superior finish.

Although old materials still play an important part in modern life they are now joined by many new materials, some of which contribute to society in ways that our forebears could not have imagined. There are just too many to discuss in a comprehensive manner so, instead, just two types, synthetic fabrics and semiconductors that are both well established, will be selected to illustrate the range of modern materials that influence our lives today. To these two will be added another category of materials that will potentially be of great importance in the future, those based on the science of extreme miniaturization — *nanotechnology*.

22.2 Manufactured Fibres, Synthetic Fibres and Plastics

Rayon

The range of natural fibres available for spinning into yarn and then weaving into cloth gave a variety of fabrics suitable for wear under all conceivable conditions. However, they varied greatly in their cost and silk, the most attractive of the fabrics, was also the most expensive and outside the financial compass of any but the wealthiest in society. Towards the end of the 19th century, in France, the first production began of a manufactured fibre designed to be a cheap alternative to silk and described by the manufacturers as 'synthetic silk'. The basic raw material for the production of this new fibre, now known as *rayon*, was cellulose, the major component of wood and the stems and harder parts of many plants. Cellulose consists of a long chain of molecules of *D-glucose*, a sugar sometimes known popularly as *grape sugar*. Part of a chain, showing the linking of two sugar molecules is shown in Figure 22.1.

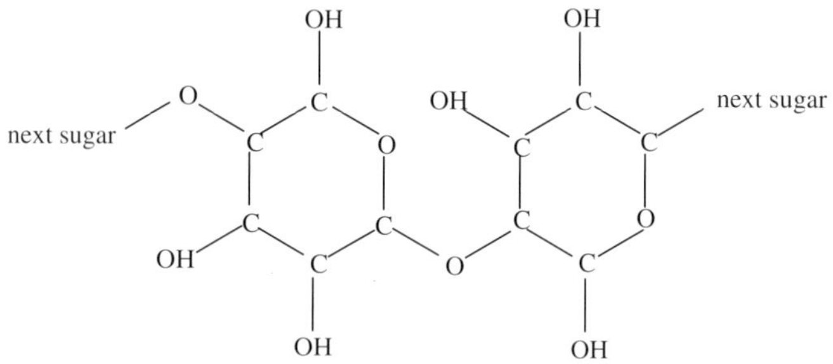

Figure 22.1 Two units of a cellulose chain.

The original forms of what is now rayon had various disadvantages, an important one of which was their high flammability. After various stages of development, involving chemists from several countries, an economic and reliable method was found for producing the rayon that we now use. The first step in the manufacturing process is to treat the cellulose with caustic soda, NaOH, which produces a pappy material known as *soda cellulose*. This is then squeezed to remove surplus liquid and then shredded to form a material known as *white crumb*. This is left for a while to oxidize, a process that leads to the formation of short polymer chains. Next the white crumb is treated with carbon disulphide, CS_2, a liquid that reacts with it to give *yellow crumb*. Finally, the yellow crumb is dissolved in a caustic solution and after being left for some time it turns into a smooth viscous liquid known as *viscose*. This is then forced through fine nozzles to create threads that finally pass though a bath of dilute sulphuric acid that reacts with the viscose and creates the cellulose filaments that constitute rayon. The thread is then stretched to align the polymer chains, which increases the strength of the filaments. These are cut into lengths of 10–15 centimetres before being spun into yarn. If the viscose is extruded through wide narrow slits before passing into the sulphuric acid then the result is *cellophane*, a transparent wrapping material in the form of sheets.

This rather elaborate process involves several chemical steps and the effluent from the production of rayon can lead to pollution of the environment. Despite the expense of producing rayon, which includes the need to avoid excessive pollution, rayon fabrics have many attractive qualities so that they are still produced, although often blended with other cheaper synthetic products. You will notice that the term 'synthetic' has not been used in relation to rayon. Rayon fibres consist of strings of cellulose molecules of the same kind as those of natural cellulose contained in wood pulp, which was the raw material for their production. The process simply turns one form of cellulose into another form so that, while rayon is certainly a *manufactured* product, it can hardly be described as synthetic.

Synthetic Fibres and Plastics

From the time of the discovery of how to produce rayon there have been many other manufactured fibres produced, but of the synthetic kind in the sense that they are of a chemical form that does not occur in nature. We just describe three of these as examples of their kind. The first two of them are of great importance to the general textile industry while the third is a specialist product with specialist applications.

Silk and wool are proteins, long chains of linked amino acids as illustrated in Figure 20.3. Cotton and linen, both vegetable fibres, have cellulose as their main constituent and, indeed, cotton is about 90% cellulose whereas wood, the main source material for rayon, is just 50% cellulose. The common characteristic of fibres, both real and artificial, is that they consist of some kind of long chain of chemical units linked by strong chemical bonds that will give the material the strength to resist tensile forces.

Nylon

The general-purpose fibre nylon was, like rayon, originally intended just as a substitute for silk. An important early use of nylon was for the production of nylon stockings to replace the expensive silk stockings that were the preceding luxury accessory; in the world of women's fashions the word '*nylons*' means nylon stockings. Wallace Carothers, a chemist who worked for the American chemical company Dupont, invented nylon in 1935. Actually nylon is a generic term since there are many different forms of nylon, all varying in their detailed chemistry while similar in type. The form of nylon invented by Carothers is known as nylon 6–6 and its chemical structure is shown in Figure 22.2. These units are joined end to end to make long polymer chains and the bonding of the units, which is similar to that in proteins, is very strong.

Nylon is produced by the chemical reaction of a *diamine* with a *dicarboxylic acid*. A diamine is a molecule where two amino acid residues, of the type illustrated in Figure 20.4 are linked together and a dicarboxylic acid is a molecule where some chemical residue, indicated

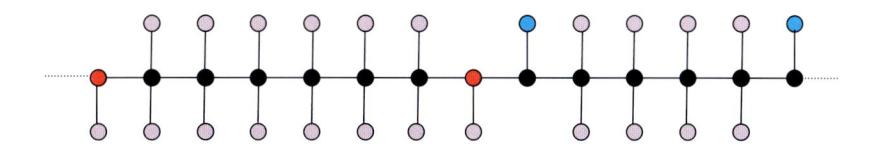

Figure 22.2 A basic unit of nylon 6–6. ● = carbon, ◉ = nitrogen, ● = oxygen, ○ = hydrogen.

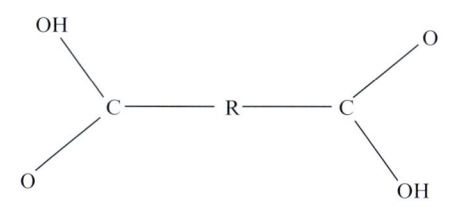

Figure 22.3 The form of structure of a dicarboxylic acid. R is a chemical residue.

as R, is sandwiched between two COOH groups as shown in Figure 22.3. The nature of the nylon produced depends on which diamine and dicarboxylic acid are used. Nylon is a thermoplastic, a material that is a solid when cold but becomes a viscous fluid when heated, at which stage it can be moulded into a variety of shapes. Solid nylon, in bulk form, has a variety of uses from making combs to making gear wheels for some applications — for example, in video recorders or other devices where large forces are not in play and silent running is desirable. Extruded into the form of filaments they form the basis for creating very strong yarns and cloth. Nylon came into its own in World War II when it was used in various military applications such as in the making of parachutes and as reinforcement in vehicle tyres.

Depending on how the nylon fibres are formed it can take on a variety of appearances when finally woven into cloth. It can have sheen, similar to that of silk, or it can have a rather dull heavy appearance. It is resistant to abrasion and is immensely strong when formed into ropes, which are favoured by mountain climbers for their combination of lightness and strength. The strength of nylon filaments is not due just to the strong chemical bonds linking the units within each chain but also due to hydrogen bonds (§20.4) that form between different chains in the filament. These occur at many points along the chains and, since they prevent the chains from moving relative to each other, it greatly increases the overall strength of the fibres.

Polyesters

The second type of general-purpose synthetic material, which gives both fibres for fabrics and bulk plastics for other uses, goes under the generic name of polyesters. Just as for nylon this substance is a polymer that can be either extruded in the form of a fibre or used in bulk form to make either sheet material or containers of various sorts. One widely used type of polyester is polyethylene terephthalate, generally known as PET, one chemical unit of which is illustrated in Figure 22.4.

Polyester cloth has many desirable properties such as its strength and resistance to shrinkage when laundered. Another useful property is that if it is exposed to flame it will scorch and may briefly catch fire but the fumes produced will then extinguish the flames. This fire resistance makes it particularly suitable for home furnishings such as curtains and sheets. Because pure polyester fabrics do not have the 'friendly' appearance and feel qualities of natural fibres they are often blended with natural fibres to give a fabric with a better visual impact — but one that still has the good qualities of polyesters.

There are many other uses of polyesters other than as fabrics. Some examples are to make bottles and electrical insulation in the form of tape or wire covering. In bulk form, reinforced with fibreglass

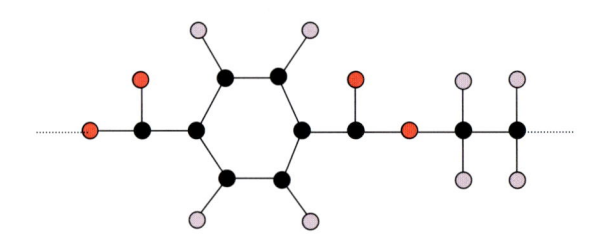

Figure 22.4   One structural unit of PET with colour symbols as in Figure 22.2.

they are used for making sailing boats or even parts of cars such as bumpers. They can also be applied in the form of a spray to cover materials and when dry they form a durable hard surface that can be sanded to smoothness and then will take a high polish.

The polymer illustrated in Figure 22.4 — PET — softens at a rather low temperature, which restricts its use for some purposes such as being filled with a hot liquid. Other forms of polyester exist with much higher softening temperatures so that, for example, in a food factory containers can be filled with hot treacle, which flows much more easily than cold.

Carbon Fibre

The final fibre we shall consider is *carbon fibre*, which is not of interest as a source of textiles for clothing or home use but does have important specialist use. Elemental carbon exists in two common natural forms with remarkably different properties — graphite, a black soft material that is the main component of pencil leads, and diamond, a clear crystalline form that is one of the hardest materials known. They differ remarkably in other properties as well — for example, carbon is an excellent conductor of electricity while diamond is a poor conductor.

Carbon fibres are extremely thin, usually in the range 5–8 microns[i] in thickness, or about one-quarter of the thickness of the finest human hair. An individual fibre is in the form of a tube and within the individual sheets of carbon forming the walls of the tube the atoms are linked in the same way as they are in graphite. There are two possible ways in which the many sheets forming the walls can be linked to their neighbours. Figure 22.5(a) shows in schematic form the cross-section of a tube formed from several cylindrical layers of carbon graphite-like sheets, each layer of which is linked to neighbouring sheets in the way that atoms are linked together in graphite (Figure 22.5(b)). In an actual carbon fibre the number of sheets forming the tube would be of the order of a few thousand. In another form of carbon fibre the structure is *turbostratic*, meaning that the

---

[i] 1 micron = 10^{-6} metre.

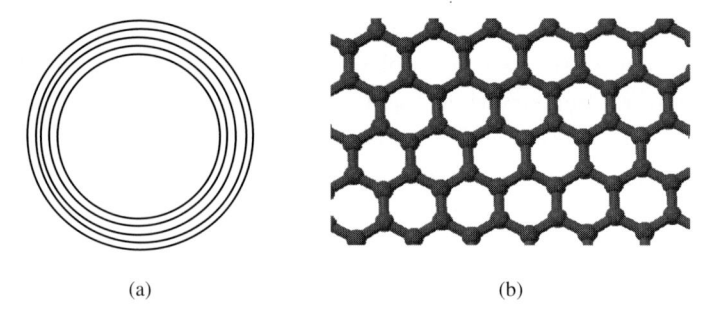

Figure 22.5 (a) A schematic cross-section of a carbon fibre in which the cylindrical layers of carbon atoms are linked as in graphite. (b) The structure of a graphite sheet with circles representing carbon atoms.

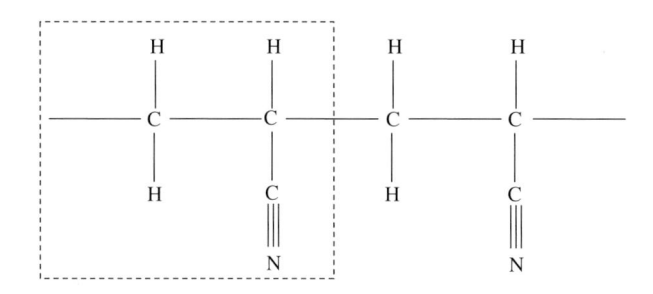

Figure 22.6 Two linked units of PAN. One unit is enclosed in a dashed box.

graphite layers are somewhat crumpled. This changes the nature of the linkages between the planes and hence the physical properties of the fibres.

The commonest starting point for the manufacture of carbon fibre is the polymeric material polyacrylonitrile (Figure 22.6), usually referred to as PAN. This is drawn into fibres, a process that ensures that the polymer chains are aligned with the fibre. These fibres are then heated in air, which breaks the hydrogen bonds linking different polymer chains and also oxidizes the material. Finally it is heated to 2 000°C in an atmosphere of argon, an inert gas, which changes the structure of the material and creates the connections that give the graphite sheets that form carbon fibre. This process of making carbon fibres gives the turbostratic form that has very high tensile strength. In tension these carbon fibres have nearly three times the strength of steel while only having one-quarter of the density. The graphitic form, produced with either pitch or rayon as a starting point, has less tensile strength but is a much better electrical conductor than the turbostratic form.

Several thousand of the very fine individual carbon fibres can be twisted together to form a yarn that can be used to make a fabric that has very high strength, good heat-resisting qualities and high electrical conductivity. These properties are useful in many applications — for example, in an industrial application where it is necessary to filter gasses at high temperature. The greatest use of carbon fibres is their incorporation into polymer plastic to make composite materials that are light and exceptionally strong. These composites find applications in the aerospace industry, the manufacture of sailing boats and as components of cars, motorbikes and bicycles — for example the competition bicycles used by Olympic cyclists. They are also to be found in many everyday items such as fishing rods, games racquets, musical instruments and computers.

22.3 Semiconductors

Try to imagine life without computers, mobile phones, iPods and all the other electronic gadgetry that fill our homes. Many older members of society did lead their early lives without such benefits, the technology for which became available during the early 1960s. This technology depends on semiconductors, materials that were first investigated by physicists who were just interested in the way that electrons moved about in solids and the role of quantum mechanics in explaining that behaviour. As in so many other cases, what was started as purely curiosity-driven research became the basis for a great industry.

Semiconductors, as their name suggests, are materials that conduct electricity — unlike insulators that do not — but do not conduct as well as most metals. The conduction of electricity through a body means that electric charges are moving through it. The rate of flow of charge through a body per unit time is known as the current that is measured in the units amperes (§7.1). The charge carriers are usually electrons and metals are good conductors because they contain many electrons that are so loosely bound to their parent atoms that they can more-or-less move freely through the metal. When an electric potential difference is applied across the conductor the electrons flow through it to create the current. At the other extreme in an insulator all the atomic electrons are so tightly bound to atoms that even the application of a large potential difference will not budge them so that no current will flow. Between these two extreme kinds of material there are semiconductors that do have some loose electrons that can flow and provide a current, but not to the same extent as in a conductor.

There is an interesting difference in the way that conduction varies with temperature for a conductor and semiconductor. Heating up a conductor does not greatly change the availability of free electrons and the main effect of temperature is to agitate the electrons so that they undergo frequent collisions and flow less easily. Consequently, heating a conductor *decreases* its conductivity. For a semiconductor the effect is quite different. With a higher temperature more electrons are able to escape from their parent atoms and this is dominant over the agitation effect. By contrast with a conductor, when a semiconductor is heated its conductivity *increases*.

The main material for the production of semiconductors is the element silicon, the electronic structure of which is illustrated in Figure 13.1. It is not the only semiconducting material — germanium, with atomic number 32, is another — but the importance of silicon is indicated by the names given to major centres of the electronics industry — Silicon Valley in California and Silicon Glen in Scotland.

Silicon is chemically related to carbon (Figure 13.1) and has a valency of four. In the crystal structure of pure silicon each atom is at the centre of a tetrahedron of other atoms to which it is bound (Figure 22.7). The electrons shared between neighbouring silicon atoms gives each of them an environment with a full 2p shell containing eight electrons and if pure silicon is at a very low temperature it will behave like an insulator since there are no free electrons to carry a current. However, if the silicon is heated then some of the electrons can be shaken loose from their bonded positions and so become

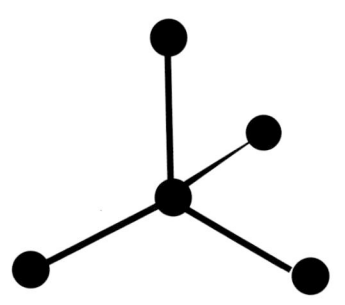

Figure 22.7 The bonding of an atom within a silicon crystal.

available to act as conducting electrons. Now we come to an interesting concept. For each electron that is free to move through the crystal there is clearly an empty space in the crystal structure where there is a gap in a complete shell. This gap is called a *hole* and behaves as though it is a particle with a positive charge. If an electrical potential difference is applied across the crystal then a hole, a position where the bonding is incomplete, will migrate across the crystal from one silicon-silicon bond to the next in the opposite direction to the movement of the electrons. Negative charge moving in one direction and positive charge moving in the opposite direction both contribute to current in the same direction. As the temperature of the silicon is increased so the number of free electrons, and the number of corresponding holes, increases so increasing the conductivity. A semiconductor material like pure silicon, containing an equal number of electrons and holes, is called an *intrinsic semiconductor*. The property that the conductivity increases with temperature is exploited in a device called a *thermistor*, one use of which is to act as a temperature control so that when it becomes too hot and the current through it rises to a certain level then the source of heating is switched off.

The semiconducting properties of silicon can be greatly affected by adding small quantities of impurity to it, a process known as *doping*. Adding a pentavalent (valency 5) impurity such as phosphorus or arsenic gives what is known as an *n-type semiconductor*. Having five electrons in its unfilled outer shell a phosphorus atom has one electron with no bonding role to play when it is inserted into the

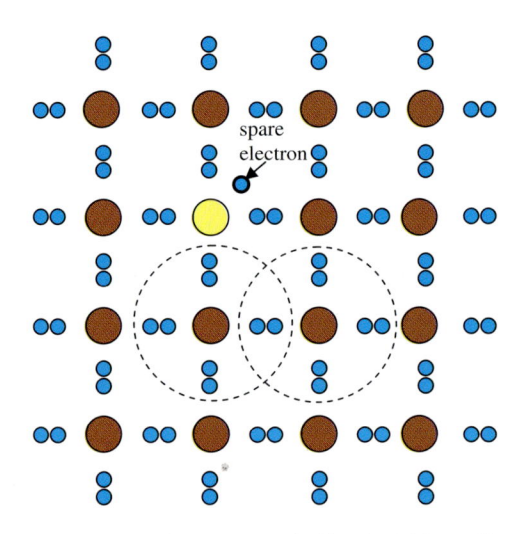

Figure 22.8 A phosphorus atom (yellow) within a silicon crystal.

structure of a silicon crystal. This situation is represented in two dimensions in Figure 22.8, although the structure is actually three-dimensional. By sharing their unfilled outer shells all the atoms have an environment that gives them a filled outer shell (within dashed circles). The phosphorus atom also enjoys a filled-shell environment but it provides an extra electron that can move freely through the crystal and so give electrical conduction.

An *n*-type semiconductor does not depend for its conducting properties on temperature — it acquires its conducting electrons from the electronic structure of the impurity. It is called *n*-type because the '*n*' stands for negative, the sign of the charge on the current carriers.

By adding a trivalent impurity, such as boron or gallium, to silicon a *p-type* semiconductor is produced. This situation is illustrated in Figure 22.9. With only three electrons in its outer shell there is a deficiency of electrons to give a full shell around all the atoms. The resultant hole, which acts like a particle of positive charge, gives the material conductivity and the '*p*' in *p*-type indicates that the effective carriers of a current have a positive charge.

Introducing charge carriers by doping gives what are called *extrinsic semiconductors*. Individually the *n*-type and *p*-type extrinsic

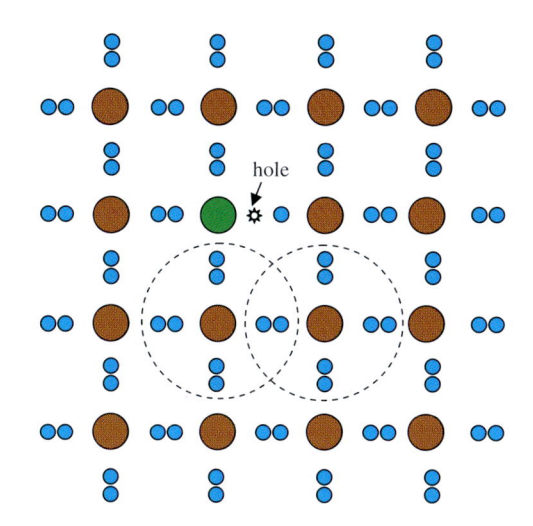

Figure 22.9 A gallium atom (green) is deficient in an electron to complete a shell.

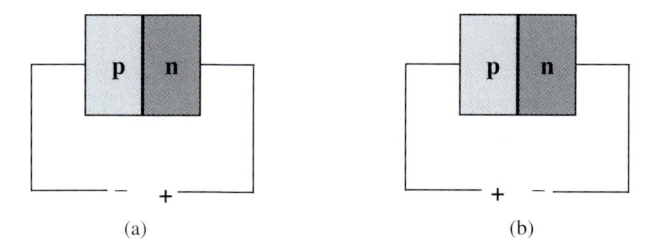

Figure 22.10 Semiconductor diode in (a) non-conducting configuration and (b) conducting configuration.

semiconductors just behave like conducting materials, that conduct electricity moderately well but not as well as most metals. However, when the two types of extrinsic semiconductors are combined in various ways they can produce devices with interesting and useful properties. One of these is illustrated in Figure 22.10. It consists of a slice of *p*-type semiconductor joined to a slice of *n*-type. If a battery is connected across the device as shown in Figure 22.10(a) then no current will flow. The negatively charged electrons are attracted towards the positive side of the battery and the holes are attracted towards the negative side of the battery. No charge carriers cross the junction between the two types of semiconductor and hence no current flows through it. However, if the battery is reversed, as shown in Figure 22.10(b), then a current will flow. The electrons in the *n*-type semiconductor are attracted towards the junction by the positive side of the battery and similarly the holes are also attracted towards the junction by the negative side of the battery. At the junction the electrons moving across the junction effectively fill the holes. However, the electronic structures of the two types of semiconductor are not changed by the passage of a current; new holes and free electrons are constantly being created within the materials to maintain the current flow.

This kind of device, which allows a current to flow one way but not the other is a *diode*, which acts like a *rectifier*. If an alternating potential difference is applied to a diode then only when the potential is in the admitting sense will a current flow through it. The direct current that passes through the device is non-uniform but can be smoothed out to give a reasonably uniform direct current.

By combining *p*-type and *n*-type semiconducting materials in various ways electronic devices with other useful properties can be produced and such devices are in abundant use at the present time.

22.4 Nanotechnology

In 1959 an outstanding American physicist Richard Feynman (1918–1988; Nobel Prize for Physics, 1965) gave a talk to the American Physical Society entitled *There's Plenty of Room at the Bottom*. In this talk he envisaged a branch of science and technology based on constructing devices that would be on the scale of small groups of atoms or molecules where the forces operating between atoms would be far more important than they usually are on the larger scale and, in particular, gravity would be irrelevant. The Japanese scientist Norio Taniguchi (1912–1999) took this idea further when in 1974 he coined the word *nanotechnology* to describe scientific activity involving entities smaller than about 100 nanometres, or 10^{-7} metres. This is a field in active development in many different areas of science; here we shall just describe two kinds of nanotechnological development.

Carbon Nanotubes

The element carbon is extremely versatile in the structures it can form — for example, very soft graphite, extremely hard diamond and strong carbon fibres. A new form of carbon, discovered by the British chemist Harry Kroto (b.1939; Nobel Prize for Chemistry, 1996) is a ball-like molecule of 60 linked carbon atoms that will all sit on the surface of a sphere. The arrangement of atoms has a strong affinity with the form of the leather panels of a football (Figure 22.11); the carbon atoms are situated at the corners of the hexagons and pentagons. This material, *Buckminsterfullerene*, was the forerunner of many other three-dimensional framework carbon structures — *the fullerenes*. These can have ellipsoidal or tubular shapes or even be planar and contain different numbers of carbon atoms — for example the ellipsoidal C_{70}.

When the individual C_{60} molecules of Buckminster Fullerene come together to form a solid the strength of the bonding between molecules is much weaker than the bonds within the molecules. For this reason it is possible to interpose doping atoms within the solid, a process known as intercalation. Some of these doped materials have interesting properties. Un-doped Buckminster Fullerene is a good conductor of electricity but the doped material, cooled to a very low temperature, can become a *superconductor* — that is it will offer no resistance at all to the passage of electricity. Doping with the metal

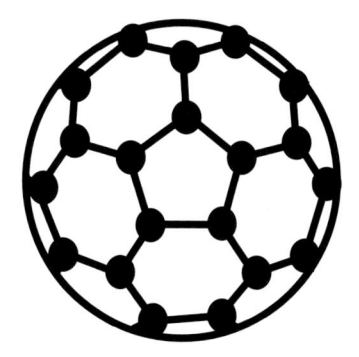

Figure 22.11 The 60 carbon atoms in Buckminsterfullerene are at the apices of the hexagons and pentagons that are arranged round a spherical surface.

caesium to give Cs_3C_{60}, the resistance falls to zero at a temperature of 38 K,[j] which is quite a high temperature for superconductivity, although there are some materials with higher transition temperatures.

The way that the fullerenes are produced is by striking an electric arc between carbon electrodes with a very high current of the order of 50 amperes. The carbon of the electrodes is vaporized and when it cools it produces some C_{60} and other fullerenes. In 1991 the Japanese physicist, Sumio Iijima (b.1939), found another form of carbon in the products from a carbon arc — *carbon nanotubes*. In §22.2 we described carbon fibres, fine tubes of concentric connected cylinders of carbon atoms that could be spun into a yarn and then woven into a fabric. The individual fibres are a few time 10^{-5} metres in thickness, a small fraction of the width of a human hair, and several thousand of these fibres spun together form a usable yarn. Carbon nanotubes differ from carbon fibres in that there is either a single cylinder of carbon atoms or very few concentric cylinders. These nanotubes have widths of a few nanometres (about 1/100 000 of the width of a human hair) and lengths of up to a few millimetres (Figure 18.7). The hexagonal carbon units that form the walls of the nanotubes can have various relationships with the direction of the axis of the tube. Representations of sections of two different forms of single-walled carbon nanotubes are shown in Figure 22.12.

The physical properties of carbon nanotubes are very remarkable and many potential uses for them are being investigated. The hexagonal structures in the walls are extremely rigid and for that reason their resistance to being stretched is five times higher than that of steel. The tensile strength is also very high, being about 50 times that of steel. These mechanical properties, allied to their low density have suggested that they could have applications in the aerospace industry.

Another interesting feature of carbon nanotubes is that, depending on their structure, they can be either conductors or semiconductors. They can be better conductors than copper, which is an exceptionally good metallic conductor, meaning that there are less power losses due

---

[j] 0 kelvin (symbol K) is the absolute zero of temperature. The intervals of temperature on the Kelvin scale are the same as those on the centigrade (Celcius) scale.

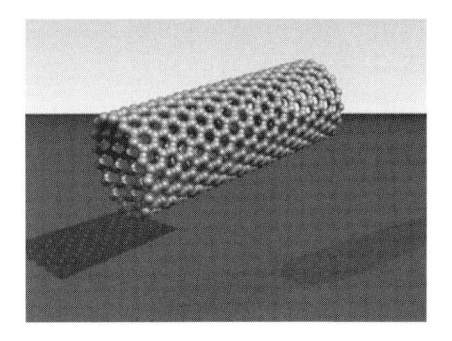

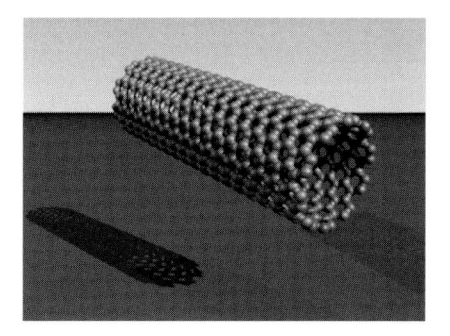

Figure 22.12 Two forms of carbon nanotube (a) zigzag (b) armchair.

to heating if currents are passed through them. The carbon nanotubes with semiconducting properties behave like silicon and there is the possibility of building electronic devices on a nanometre scale if the problem of handling such small entities can be solved.

Rotaxanes

The next type of nanotechnology device is described by the name *rotaxanes*. A rotaxane is a molecular device that can be represent in the diagrammatic form shown in Figure 22.13. The component marked A represents a ring-like molecule, which could be something like several six-membered rings joined together to form a complete loop. The remaining part of the rotaxane is a dumb-bell shaped molecule, threaded through the ring with the structures at each end of the 'bar' larger than the space enclosed by A. It is clear that the parts of the structure, A and B, cannot be separated without disrupting one or other of the parts.

Although a rotaxane structure cannot be disrupted without breaking chemical bonds in some part of it, there can be relative motion of the dumb-bell and the ring without disruption. For example, the ring can rotate around the dumb-bell axis or slide along the axis towards one or other of the two ends. Such movements can be made to happen by a chemical reaction, by light or by applying an electric field. Two different positions of the ring relative to the dumb-bell would constitute a bipolar switch — like a light switch that is either on or off — and

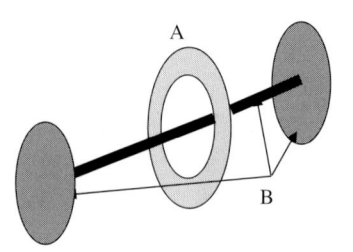

Figure 22.13   A diagrammatic representation of a rotaxane.

this is the basis of how a computer operates in binary mode with its basic memory elements representing either 0 or 1. There are many potential uses for rotaxanes if the bipolar-switch property can be readily and reliably detected and exploited.

Nanotechnology is likely to provide the next major jump in technological advancement, although when is still not certain. It has had a bad press in some circles where it has been presented as a potential threat to the world. In 1986 an American scientist, Eric Drexler, in a book *Engines of Creation* envisaged the production of self-replicating nanotechnology devices that would consume all the available material in the world and turn it into a 'grey goo'. This and other unlikely Doomsday scenarios have occasionally been presented in popular science writing but there is really no justification for any alarm connected with this new advance in technology.

Chapter 23

The Fantastic World of Particles

23.1 Antics on Ice

The structure of an atom, established in the early years of the 20th century, was that it consisted of a central nucleus, containing of a tight collection of neutrons and protons within a space of dimension about 10^{-15} metre, surrounded by electrons at distances of order 10^{-10} metre, with electrostatic forces of attraction between the positively charged nuclei and the negatively charged electrons. On the basis of just the existence of electrostatic forces it seems impossible for a nucleus to exist. Neutrons are electrically neutral but the positively charged protons in the nucleus are close together, so the repulsive electrostatic forces on them should be enormous thus causing the nucleus to fly apart. The fact that it does not do so indicates the existence of some other very strong force binding the nucleons (protons and neutrons) together.

One of the great concepts that Isaac Newton introduced was the idea of force at a distance without the intervention of material linking the bodies between which the force acts. This is the way that the gravitational force operates and the vacuum of space is no bar to its operation. Now, by contrast, we look at a model for a force between bodies that *does* involve some material interaction between them and later we shall relate this model to the world of the fundamental particles that constitute the whole of the material world.

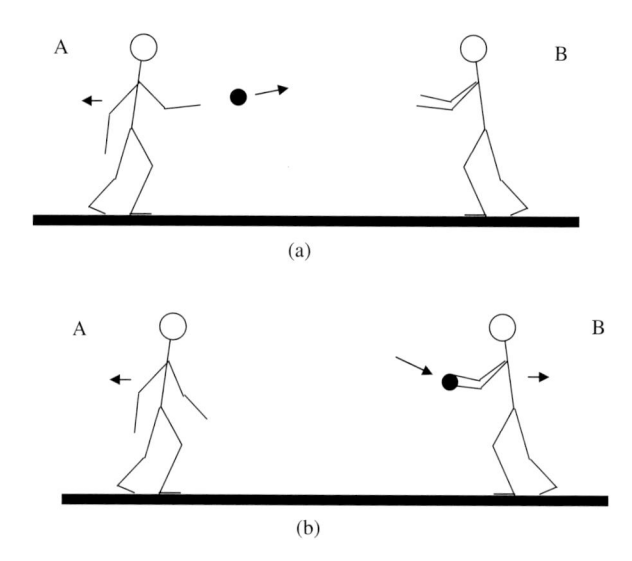

Figure 23.1 (a) A throws a ball towards B and is propelled backwards away from B
(b) B catches the ball and is propelled backwards away from A.

Consider a situation where there are two people standing on a smooth sheet of ice (Figure 23.1). They start at rest but then person A throws a ball to person B. At the instant the ball is thrown towards B there is a reaction that propels A backwards away from B. Now B catches the ball and the momentum of the ball propels B backwards away from A. We can see that if the ball is thrown to-and-fro between A and B the net effect is that A and B move away from each other, just as though there was some force of repulsion between them. Now let us suppose that the ball they were throwing was invisible. The fact that there was some material interaction between the two people would not be evident but what would be seen is that they were moving apart and it would be concluded that there was some repulsive force at work.

To push the model a little further so as to include an attractive force we can link the two individuals by throwing a boomerang. This is an Australian aboriginal throwing stick used for hunting and its main purpose is to kill or stun the hunted animal rather than to perform a curved path through the air. There is also a misconception

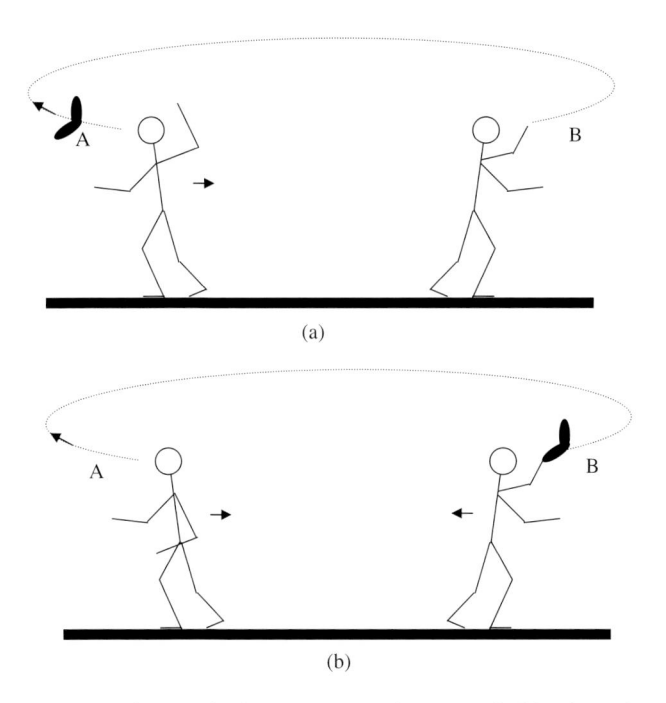

Figure 23.2 (a) A throws the boomerang and is propelled backwards towards B.
(b) B catches the boomerang and is propelled backwards towards A.

that the boomerang can be thrown so as to return to the thrower retracing its outward path, but this is not so. The boomerang actually follows a more-or-less circular or elliptical path. We show the effect of throwing a boomerang between two people in Figure 23.2.

When A throws the boomerang it moves in the opposite direction to B and hence nudges A backwards towards B. At the time the boomerang is caught by B it has performed a circular motion that, if B had not caught it, would have returned it to A. The impetus given by the boomerang when caught propels B towards A. If the boomerang were being thrown to-and-fro between A and B then the net effect would be to steadily push A and B towards each other. However, if the boomerang were invisible then the motions of A and B would be interpreted as being due to some attractive force between them.

In 1935 the Japanese physicist, Hideki Yukawa (1907–1981) created a theory for the strong force attracting nucleons to each other that postulated a particle playing the role for two nucleons that the boomerang played for the individuals A and B. From the fact that the strong force acted only over tiny distances, corresponding to the size of the nucleus or less, Yukawa was able to estimate the mass of the hypothetical particle as about 200 times the mass of an electron and from general theoretical considerations he found that it had to be a boson — that is it had to have integral spin (§14.3). Because its mass is intermediate between that of an electron and a proton it was called a *meson*, a name of Greek origin implying 'something between.' In 1947 a British physicist, Cecil Powell (1903–1969) was carrying out experiments in which cosmic rays produced tracks on photographic plates carried to high altitudes on balloons. From the nature of the tracks the characteristics of the particles could be assessed. For one track a mass 207 times that of the electron was found that seemed to fit in with Yukawa's prediction. Yukawa and Powell were both awarded the Nobel Prize for Physics, in 1947 and 1950 respectively, although, as it turned out, the particle found by Powell was not the one predicted by Yukawa. Powell's discovery was a *fermion*, a particle of half-integral spin and hence could not have been Yukawa's meson. Later, three particles were discovered that did match Yukawa's prediction. They were bosons, had approximately the predicted mass and differed only in that one possessed a positive charge, one a negative charge and the third one no charge. These particles are called *pi-mesons* or, more commonly *pions*. The particle found by Powell is called a *muon*, a negatively charged fermion that behaves rather like a heavy electron, except that it is not stable and decays with a half-life of just over 2 microseconds.

The earliest work in discovering new exotic particles was based on the Powell technique of using cosmic rays as a natural source of high-energy particles that would occasionally collide with an atom on the photographic plate and release enough energy to create some new particles. However, it was a rather hit-and-miss technique, dependent on chance events, and something more controllable would obviously be an advantage.

23.2 The First Designed Atom-Smasher

The first recognized and recorded atomic disintegration was the experiment by Rutherford in 1919 when an α-particle from a radioactive source interacted with a nitrogen nucleus to give a proton plus an oxygen nucleus (§10.3). The energy of the α-particle was several MeV (million electron volts) and clearly energies of this order of magnitude were required if nuclei were to be disrupted. In 1932, at the Cavendish Laboratory, Cambridge, the British physicist, John Cockroft (1897–1967; Figure 23.3) and his Irish colleague, Ernest Walton (1903–1995; Figure 23.3) succeeded in accelerating protons to an energy at which they could disintegrate some nuclei with which they collided. The secret of their success was that they developed an ingenious way of producing a source of very high potential difference through which the protons could be accelerated — the so-called Cockroft–Walton generator. By repeatedly doubling an initially relatively low voltage they were able to produce a source of potential difference of more than 700 000 volts.

The Cockroft–Walton experiment, in schematic form, is shown in Figure 23.4. An electric discharge through hydrogen produces protons, which then pass through a hole in plate A and are accelerated towards plate B at a potential 710 000 volts lower than that of plate A. After

Figure 23.3 John Cockroft (left) and Ernest Walton (right; courtesy Trinity College, Dublin).

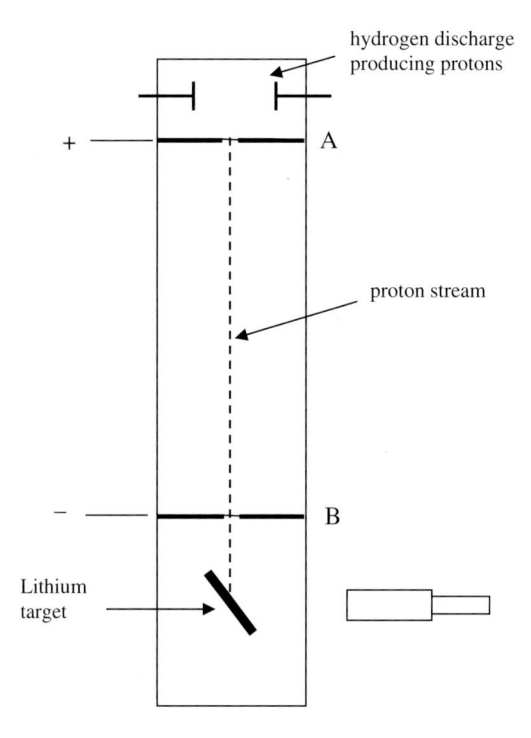

Figure 23.4 The Cockroft–Walton apparatus in schematic form.

passing through a hole in plate B the fast protons impinge on a target, the nuclei of which are to be disrupted. With a target of lithium the following reaction was found to have taken place

$$_1^1p + {}_3^7Li \rightarrow 2\,{}_2^4He.$$

Cockroft and Walton were able to detect the presence of the helium nuclei and also found that energy was being produced in the process. For this work they were jointly awarded the Nobel Prize in Physics in 1951.

The products of the reaction just described had less mass than the original particles and the loss of mass was transformed into energy in conformity with Einstein's equation $E = mc^2$, the first practical demonstration of that formula. This experiment, in common with other experiments of a similar kind conducted shortly afterwards, used more energy in carrying out the experiment than was produced by the transformation of mass to energy. This led Rutherford to make the following comment in an article in the journal *Nature*: 'These transformations of the atom are of extraordinary interest to scientists but we cannot control atomic energy to an extent which would be of any value commercially, and I believe that we are not likely to be able to do so. Our interest in the matter is purely scientific, and the experiments which are being carried out will help us to a better understanding of the structure of matter.' Even great scientists can sometimes be wrong!

23.3 The Cyclotron

At about the same time that Cockroft and Walton carried out their seminal experiments, the American physicist Ernest Lawrence (1901–1958; Nobel Prize for Physics, 1939) at the University of California, Berkley, designed a new kind of particle accelerator, the *cyclotron*, shown in schematic form in Figure 23.5.

Charged particles are injected at the centre of the cyclotron and are accelerated in the gap between the two D-shaped electrodes that have an alternating potential difference between them. A magnetic

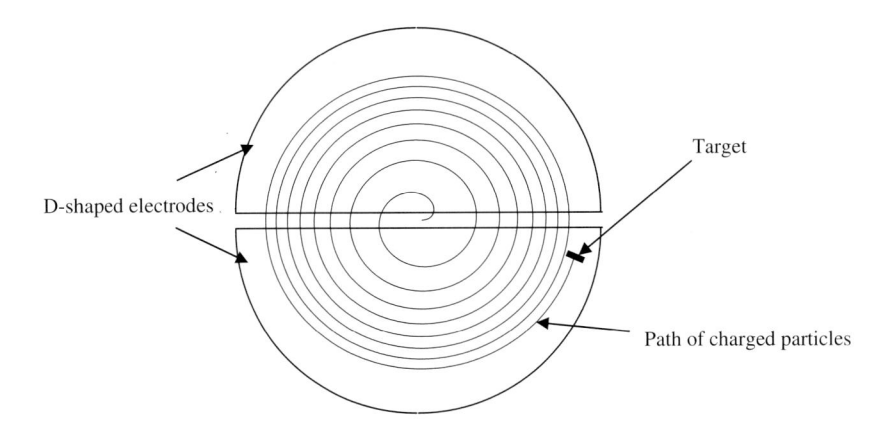

Figure 23.5 The Cyclotron.

field is applied perpendicular to the plane of the illustration and this curves the path of the charged particles. If the particles always moved at the same speed then their path would be a circle but the alternating potential is such that whenever they enter the space between the electrodes they are accelerated. Because of their increasing speed (energy) their actual path is in the form of a spiral and at the end of the path, when the energy is a maximum they collide with a target, producing new particles that can be analysed by a detector.

For many decades the cyclotron was the main tool of the particle physicist. Even today they still have some medical use as generators of beams of high-energy protons that can be used in cancer treatment.

23.4 The Stanford Linear Accelerator

There is an obvious limit to the energy of particles that can be produced by a Cockroft–Walton apparatus, dictated by the potential difference that can be maintained between the plates A and B. A device called a linear accelerator (LINAC) overcomes this limitation. The largest and most powerful of these is the Stanford linear accelerator, which is in the form of a tube 3.2 kilometres long. The LINAC uses a completely different method of applying accelerations to charged particles that negates the need for generating enormous potential differences. The basic principle that operates to accelerate the particles is illustrated in Figure 23.6.

In the figure the grey circle represents a tight bunch of charged particles, which could be electrons, positrons, protons or some heavier ions. Let us take the particles as electrons for this description. When the electrons reach the gap between copper cylinders A and B

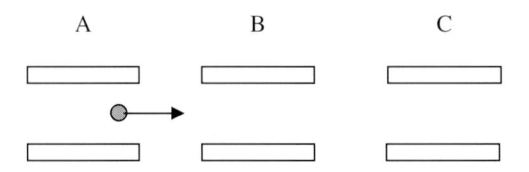

Figure 23.6 A basic component of a linear accelerator. A, B and C are copper cylinders and the grey circle a bunch of charged particles.

the potential difference between the cylinders is such as to accelerate the electrons within the gap — which requires the potential of B to be higher than that of A. While the electrons are moving through cylinder B they will be in a constant potential and so will not be accelerating. However, while they are moving through B the potential difference between cylinders B and C is altered so that by the time the electrons reach the end of B the potential difference between B and C is such that they are further accelerated while travelling in the gap between these two cylinders. If this process is repeated many times then the electrons can be continuously accelerated while at the same time excessively large potential differences are never required between neighbouring cylinders. The total accelerating potential along the whole system is the sum of the potential differences, occurring at different times between neighbouring pairs of cylinders, and in the case of the Stanford linear accelerator this amounts to 30 GV (1 GV = 10^9 volts). Once the particles reach the end of the LINAC they can be aimed at some chosen target and the resultant particles from the collisions can then be detected and analysed.

It is necessary for the charged particles in a LINAC to be in the form of a stream of compact bunches. The accelerating conditions can be simultaneously imposed on separated bunches in the tube but could not be imposed at one time at all points of a continuous stream of particles. An interesting characteristic of a LINAC is that it can simultaneously accelerate bunches of electrons and positrons, which can be separated along the tube so that electrons are being accelerated in a gap with an increasing potential while positrons are accelerating in a nearby gap with a decreasing potential. At the end of the tube the streams of electrons and positrons can be separated by magnetic fields and then made to collide with each other, giving annihilation and the production of γ rays.

23.5 Synchrotrons and the Large Hadron Collider

The cyclotron uses a constant frequency electric field and a constant magnetic field, which causes the charged particles to move along a spiral path. In a *synchrotron* particles are made to move on a closed

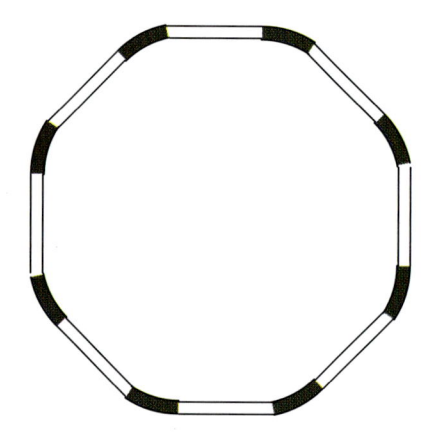

Figure 23.7 A schematic synchrotron. The black sections are bending magnets.

path that they can go round repeatedly while being accelerated up to some maximum energy. A simplified illustration of the form of a synchrotron is given in Figure 23.7.

In a synchrotron the charged particles travel around a track in an evacuated tube along several straight sections connected by bending magnets that change the direction of the particles so that they go smoothly from one straight section to another. Within the straight sections there are devices for increasing the energy of the particles. These can either be of the type previously described where the particles move through a potential difference or, more usually, they can be accelerated by microwave cavities where the charged particles are accelerated by moving with the microwaves much as a surfer rides a wave on the sea. To get the process started charged particles are inserted at high energy — from a linear accelerator, for example — and then ramped up in energy as they repeatedly traverse the ring. The 'synchro' element in the word 'synchrotron' implies that the magnetic and electric fields are synchronized with the travelling bunches of charged particles so as always to be injecting energy and keeping them within the track. The limit to the energy that can be achieved in a synchrotron is dictated by the fact that whenever the particles change direction due to the action of the bending magnets they emit radiation and the limiting energy is reached when the energy added by the microwave cavities equals the energy of the emitted radiation. The emitted radiation, which covers a wide range of wavelengths from the infrared to X-rays, is a waste product of particle physics but it has a very high intensity and is useful for a number of scientific applications, notably for X-ray crystallography (§20.1). X-ray crystallographers and other scientists first used this *synchrotron radiation* as parasitic users of machines being used for particle-collision experiments. However, the radiation was so useful that synchrotrons were built specifically to run as *storage rings* where electrons are maintained in orbit around the rings for long periods of hours or even days, emitting radiation from each of the bending magnets that may then be used for extended data collection for X-ray crystallography or other kinds of experiments.

When used for particle physics it is possible to have two beams of particles travelling round the synchrotron in different directions and in non-intersecting paths. When the desired energies are reached, they can be deflected so as to collide; various detectors can then be used to detect the resultant particles from the violent collision. Figure 23.8 shows part of the ring of the Relativistic Heavy Ion Collider (RHIC) at the Brooklyn National Laboratory, New York. Figure 23.9 is an image from one of the RHIC detectors showing the tracks of thousands of particles produced by the collision of gold ions.

Figure 23.8 The Relativistic Heavy Ion Collider (Brookhaven National Laboratory).

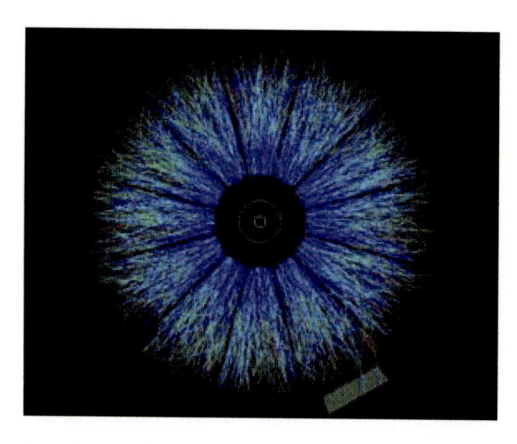

Figure 23.9 Particle tracks from the collision of very high-energy gold ions.

The most powerful machine of this kind to be built is the *Large Hadron Collider* (LHC), which is 27 kilometres in circumference and spans the Swiss/French border near Geneva. It has 9 300 superconducting bending magnets. These produce their magnetism by passing electric currents through coils of wire cooled to liquid-helium temperature (1.9 K), at which temperature the wires have zero resistance to the passage of current. Counter-rotating beams of protons or heavy particles can be produced and then caused to collide when their speeds are 99.999 999% of the speed of light. The conditions of the collision will be so energetic that it is hoped that the particles being produced will be of the type that existed in the early Universe when the energy density was extremely high. The facility is a tremendous feat of engineering and it is curious that in order to observe very tiny and elusive particles one must create experiments on such a vast scale. The scale is evident in Figure 23.10 that shows the installation of a vacuum tank that is to be part of a muon detector.

On the basis of particle-physics experiments a large number of particles have been discovered with different masses, charges and other properties. These particles, such as the aforementioned mesons, have very brief existences but their discovery disturbed the nice cosy picture of a few fundamental particles — the proton, neutron,

Figure 23.10 The installation of part of a muon detector for the LHC.

electron and, reluctantly and of necessity, the neutrino that could explain all the matter in the Universe. Clearly, a new picture had to be formulated, one that will now be described.

23.6 Some Fundamental Particles

In Chapter 16 the existence of the *antiparticles* of both the electron and the neutrino were mentioned. In the case of the electron the antiparticle, the positron, is readily observable since it is a product of some radioactive decays. It has a charge of opposite sign to that of the electron and when an electron and positron meet they spontaneously destroy each other with their mass converted to electromagnetic radiation. Although neutrinos and antineutrinos have no charge, a meeting of the two would also lead to their mutual destruction and the creation of electromagnetic radiation.

High-energy physics experiments have shown that for every particle there is an antiparticle. The product of some particle-collision experiments includes *antiprotons*, particles with the mass of a proton but with a negative charge. The presence of *antineutrons* has also been inferred in some experiments. Since these particles, like neutrons, have no charge it is difficult directly to identify them but if they happen to interact with normal matter then the product of such an interaction can reveal that they were involved. In particle physics experiments there are very many different particles that have been identified; we have already mentioned the muon, the fermion discovered by Powell, and the meson, a boson predicted by Yukawa of which three varieties were subsequently discovered. An analysis of these particles, and the interactions they undergo has led particle physicists to the identification of a reasonably small number of *fundamental particles*, combinations of which can explain all the myriad of other particles that have been observed. These are all identified by their masses, charges and spin; those with half-integral spin are fermions and those with integral spin (including zero) are bosons.

Leptons

There are three fundamental particles, in a category known as *leptons* (derived from the Greek word meaning 'thin'), which have a family relationship. They are all fermions, with ½ spin, and all have a negative charge equal to that of the electron. One of these particles is the electron itself, another is the muon, discovered by Powell with a mass 207 times that of the electron and the third is the *tau particle* with a mass some 3477 times that of the electron, or almost twice the mass of a proton. There are three other particles designated as leptons — three kinds of neutrinos. They all have zero charge and extremely tiny masses — so small that before sensitive experiments were done to estimate neutrino masses it was thought that they might have zero mass. The neutrino and antineutrino referred to in §16.4 are actually the *electron neutrino* and *electron antineutrino* because they are produced in association with either an electron or a positron (anti-electron) when neutrons decay or a proton disintegrates. The neutrinos that are produced in company with muons and tau particles are referred to respectively as the *muon neutrino* and the *tau neutrino* respectively, and the corresponding antineutrinos exist corresponding to the antimuon and antitau particles.

The six leptons that have just been described are illustrated in Figure 23.11 together with their masses given in units of the electron mass.

1	207	3477
electron	muon	tau
electron neutrino	muon neutrino	tau neutrino

Figure 23.11   The six leptons.

Quarks and Their Products

The leptons that have just been described are *fundamental* elementary particles that it is believed cannot be further decomposed into even more basic constituents. However, there are many other exotic particles that can be produced in high-energy experiments — *lamda, omega* and *kaon particles*, for example, which, together with the proton and neutron and their antiparticles, can be decomposed into even more basic constituents. Since 1964 there has been developed a new model of the structure of these kinds of particles involving a basic set of constituent particles called *quarks*. The idea of quarks was suggested independently by the American physicists Murray Gell-Mann (b.1929; Nobel Prize in Physics, 1969) and George Zweig (b.1937). There are six types of quark whose different characteristics are described as their *flavours* and are individually given the names *up*(u), *down*(d), *strange*(s), *charm*(c), *bottom*(b) and *top*(t). They are all fermions, with spin $\frac{1}{2}$. The charges associated with quarks are multiples of $\frac{1}{3}$ of an electronic charge, either $-\frac{1}{3}$ or $+\frac{2}{3}$, and the charges associated with the various quarks, as fractions of an electronic charge, are:

$$u^{\frac{2}{3}} \quad d^{-\frac{1}{3}} \quad s^{-\frac{1}{3}} \quad c^{\frac{2}{3}} \quad t^{\frac{2}{3}} \quad b^{-\frac{1}{3}}$$

There are also *antiquarks* corresponding to each of these particles, indicated by the bar over the letter symbol, with opposite charges — i.e.:

$$\bar{u}^{-\frac{2}{3}} \quad \bar{d}^{\frac{1}{3}} \quad \bar{s}^{\frac{1}{3}} \quad \bar{c}^{-\frac{2}{3}} \quad \bar{t}^{-\frac{2}{3}} \quad \bar{b}^{\frac{1}{3}}$$

Combinations of three quarks of the up-down variety form protons and neutrons. Thus

u + u + d has a charge, in electron units, $\frac{2}{3}+\frac{2}{3}-\frac{1}{3}=1$ and is a proton

$\bar{u}+\bar{u}+\bar{d}$ has a charge, in electron units, $-\frac{2}{3}-\frac{2}{3}+\frac{1}{3}=-1$ and is an antiproton

d + d + u has a charge, in electron units, $-\frac{1}{3}-\frac{1}{3}+\frac{2}{3}=0$ and is a neutron

$\bar{d}+\bar{d}+\bar{u}$ has a charge, in electron units, $\frac{1}{3}+\frac{1}{3}-\frac{2}{3}=0$ and is an antineutron

The six quarks, with their charges and masses (both in electron units) are shown in Figure 23.12.

There have been disputed claims of the observation of quarks but the large number of observed particles found can be explained by combining together, usually either in pairs or in sets of three, the six quarks and the corresponding antiquarks. The proton and neutron are *baryons*, particles consisting of combinations of three quarks; other baryons are lambda particles the three types of which have compositions

$$\Lambda^0 \quad u+d+s; \quad \Lambda_c^+ \quad u+d+c; \quad \Lambda_b^0 \quad u+d+b$$

and there are also lambda antiparticles composed of the corresponding antiquarks. By adding the quark spins it will be seen that the spins of all the lambda particles are half-integral so that, like protons and neutrons, they are fermions.

$\frac{2}{3}$ 4.7	$\frac{2}{3}$ 2485	$\frac{2}{3}$ 3.35×10^5
up	charm	top
-$\frac{1}{3}$ 9.4	-$\frac{1}{3}$ 204	-$\frac{1}{3}$ 8219
down	strange	bottom

Figure 23.12 The six quarks with their charges and masses (bold) in electron units.

In contrast to baryons, quarks can come together in even numbers, usually two or four, to form various kinds of meson. Because of the even number of quarks they must be bosons but their properties, including mass and charge, will depend on the nature of the constituent quarks. For example, *kaons*, also called K-mesons, are formed from combinations of pairs of quarks one of which must be the strange quark (or antiquark) and the other either an up or down quark (or antiquark).

There are large numbers of combinations of quarks that can give rise to many families of particles but the examples given here should illustrate the general pattern of their composition.

Interaction Particles

We began this chapter by describing a model in which a force between two people could be due to the exchange of some object passing to-and-fro between them — either a ball or a boomerang. This idea is carried over into physics in various ways. For example, there are electrostatic forces that attract unlike electric charges and repel similar charges and the photon, the particle of light described in §9.2, is the exchange particle for this kind of force field. The force of gravity has a similar inverse-square law associated with it although, unlike electrostatic forces, it is always attractive and it is postulated that there is an exchange particle associated with this force called the *graviton*.

Many particles, including the neutron and proton, are explained as combinations of quarks and the exchange particle that explains the force that binds quarks together is called the *gluon*. There are other particles associated with interactions, called the *W and Z bosons*, which are extraordinarily heavy, having about 100 times the mass of a proton. Particle physicists predicted their existence as being intermediate particles in the process of β-decay, and one of the triumphs of particle physics theory was the discovery of these particles in experiments at CERN (Centre Européenne pour la Recherche Nucléaire) in 1973. However, still undiscovered is the so-called Higgs boson, a particle that would explain how some particles acquire mass. The discovery of this hypothesized particle is one of the primary objectives of Large

Hadron Collider experiments and, if it were discovered, it would round off the inventory of particles required to explain the present observations of high-energy physics.

    In conclusion it must be said that there are many theoretical physicists who feel that the present state of particle physics is unsatisfactory and is far too complicated. We have seen how ideas about the nature of matter have fluctuated between ones in which the number of components have been small — the four elements and the nuclear structure of the atom — and larger — for example, the elements of the periodic table. The passage from the nuclear atom, with its protons, neutrons and electrons, to the present number of fundamental particles is a move from simplicity to complexity so there are some who believe that behind the current complex edifice there lurks a simpler structure involving fewer fundamental particles.

Chapter 24

How Matter Began

24.1 Wailing Sirens

We are all familiar with the changes in the sound of the siren of an emergency vehicle when first it approaches and then it recedes. As it approaches its siren is heard at a higher pitch, or frequency, than it would be if it were at rest. Conversely, as it recedes the note is at a lower pitch; the change of pitch is quite rapid as it passes through its point of closest approach. This phenomenon is known as a Doppler shift, first explained and described by the Austrian physicist, Christian Doppler (1803–1853). According to Doppler's equation for describing the frequency shift, if the vehicle approaches at 10% of the speed of sound (i.e. at about 120 kilometres per hour) then the frequency heard will be increased by 10% over that for the siren at rest and, similarly, if the vehicle retreats at 10% of the speed of sound the frequency heard will be 10% less. Another way, other than frequency, of characterizing a wave is by its wavelength that is related to frequency by the relationship

$$\text{frequency} \times \text{wavelength} = \text{speed of sound}$$

and since the speed of sound is a constant this means that an increase of frequency gives a decrease in wavelength and *vice versa*. The Doppler effect in terms of wavelength is illustrated in Figure 24.1, where the distance between each arc is one wavelength.

274   *Materials, Matter and Particles*

higher pitch   lower pitch

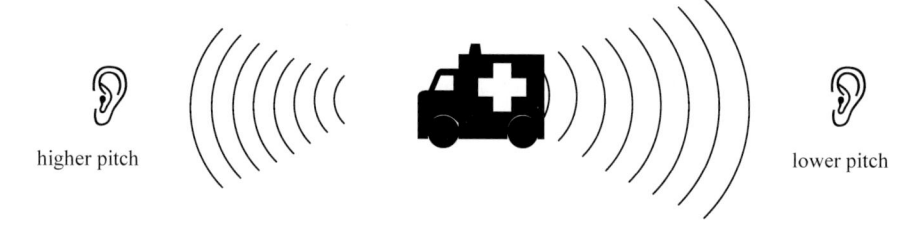

Figure 24.1   The Doppler effect. Sound waves are compressed ahead of the motion and stretched behind.

The Doppler effect applies not only to sources of sound but also to sources of any other kind of wave disturbance — which includes light. However, the speed of light is so great — 300 000 kilometres per second — that no light source on Earth moves fast enough to give an appreciable Doppler shift in wavelength. A defence in court that you went through a red traffic light because it appeared to be green would open you to a charge of speeding — at about 2×10^8 kilometres per hour!

If we move away from the Earth then we can find sources that move fast enough to be able to detect the Doppler shift in the light that comes from them by the use of scientific instruments called spectrometers. For example, stars close to the Sun typically move at speeds of 30 kilometres per second relative to the Sun, or one ten-thousandth of the speed of light. The shift in wavelength due to this speed would not be evident visually and an overall shift in the spectrum by such a small fraction of the wavelengths would not be easy to detect even with instruments. Fortunately nature gives a hand in enabling us to measure changes of wavelength corresponding to such speeds. If we examine the spectrum of the Sun we see that it is covered with fine black lines called *Fraunhofer lines* (Figure 24.2). These are caused by the absorption of light in the very diffuse outer layers of the Sun due to the presence there of different atoms in various states of ionization. Each of the lines can be associated with a particular kind of atom and if those atoms exist on other stars then the corresponding lines will appear in their spectra as well. Because these lines are so fine and so well defined, shifts in their wavelength due to motions of the stars

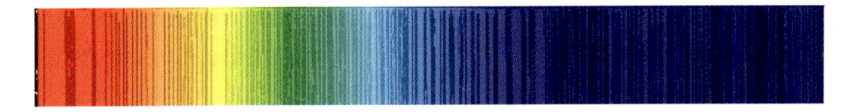

Figure 24.2 The solar spectrum showing Fraunhofer lines.

relative to the Sun (actually the Earth but this can be allowed for) can easily be measured. So precise are these measurements that variations of speeds of a few metres per second can be detected for nearby stars, for which the Fraunhofer lines are seen very clearly, and such measurements can be used to detect the presence of planets around these stars.[k]

The Doppler effect has been one of the most important tools used by astronomers in their exploration of the Universe. When stars are moving away from the Earth then the measured wavelengths of the spectral lines are shifted towards the red end of the spectrum (red-shifted). Conversely, for stars moving towards the Earth wavelengths are blue-shifted.

24.2 Measuring the Universe

Astronomers have devised various ways of measuring the distance of astronomical objects. By employing a sequence of techniques, estimates have been found for objects of ever-increasing distance. A selection of these techniques will now briefly be described.

Nearby Stars

If a relatively close star is observed at times six months apart then it will appear to have moved against the background of very distant stars. This is because the viewing point has moved by a distance equal to the diameter of the Earth's orbit; a similar effect is observed if a finger is held vertically at arm's length and then viewed first by one eye and

---

[k] This is described fully in the author's book *The Formation of the Solar System: theories old and new*.

then by the other, when the position of the finger relative to distant objects is seen to change. The shift in observed position in the distant-star background can be used to estimate the distance of the close star.

A convenient unit for measuring the distance of astronomical objects is the *light year*, the distance that light travels in one year, 9.5×10^{12} kilometres. The method just described, the *parallax method*, is capable of measuring the distance of stars with reasonable accuracy up to a distance of about 100 light years by ground observations and up to 600 light years by measurements from Hipparchos, a satellite launched by the European Space Agency in 1989 specifically to make such measurements. By this means, the distances of one million of the nearest stars to the Sun have been measured.

Measurements from Variable Stars

Some stars are found to fluctuate in brightness and one kind of star that does so is known as a *Cepheid variable*. These stars are expanding and contracting in a periodic way, in a kind of breathing mode; when contracted they are brighter than when expanded. There are many such stars within the parallax range, so their distances are known, and hence, from their apparent brightness as seen from Earth, which depends on their distance (apparent brightness decreases with distance), their *luminosity*, or total light output, can be estimated. In 1908 a Harvard astronomer, Henrietta Leavitt, found that there was a relationship between the maximum luminosity of a Cepheid variable and its period (Figure 24.3).

It will be seen from the figure that some Cepheid variables are very bright, with up to 30 000 times the luminosity of the Sun so they can be seen at great distances. Some occur in the outer parts of distant galaxies; from their periods their luminosities can be estimated and then, from their apparent brightness as seen from Earth, their distances can be found. In this way the distances of galaxies can be estimated out to 80 million light years — a long way but far from the limits of the observable Universe.

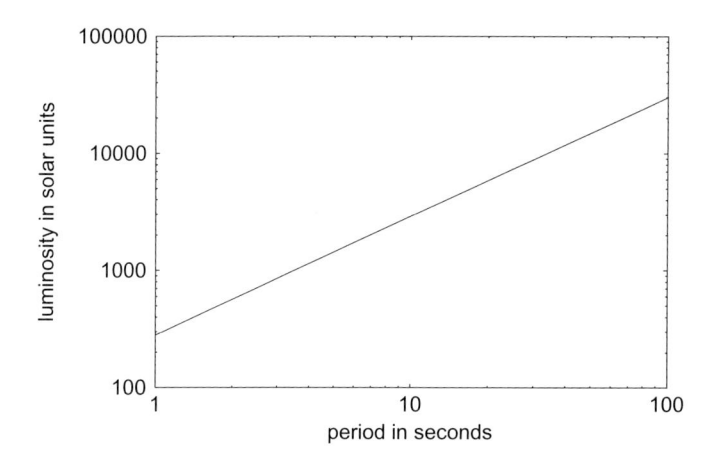

Figure 24.3 Relationship between the luminosity of a Cepheid variable and its period.

Type 1a Supernovae

The Sun is a star on the *main sequence*, that is to say that it is generating energy by nuclear processes that convert hydrogen to helium. The Sun is about 4.5 billion years old, half way through its main-sequence lifetime. Eventually it will run short of hydrogen in its central regions and will then go through a number of evolutionary changes. It will at first swell up to a size where it will just about engulf the Earth, at the same time cooling and hence becoming a red colour. At this stage it will be a *red giant*. It will then go through a stage of shedding outer material and eventually will settle down as a bright very compact object called a *white dwarf*. A white dwarf has about the mass of the Sun and a radius similar to that of the Earth so its density is about 2×10^9 kilograms per cubic metre, more than 300 000 times that of the Earth. For a star somewhat more massive than the Sun the evolutionary process is more violent and at one stage the star explodes violently to give a *supernova* when, for a period of a few weeks, it can be as bright as 10 billion Suns.

There are other ways in which violent supernovae outbursts can occur but the source of a supernova can usually be recognized by the form of its spectrum. There is one kind of supernova that originates in a binary star system, where there are two stars orbiting around each other. The kind of binary system of interest is where one star is a red giant and the other is a white dwarf. The red giant is shedding material and some of it will attach itself to the white dwarf. A white dwarf is a very stable kind of star but if it gains material and reaches a critical mass called the *Chandrasekhar limit*, about 1.4 times the mass of the Sun, then it becomes unstable and explodes to give what is called a type 1a supernova. Because these always occur under very similar circumstances, when the white dwarf reaches the same critical mass, all type 1a supernovae have the same peak brightness.

Some type 1a supernovae are observed in galaxies within the Cepheid variable range so their distances and hence their peak luminosities can be estimated. Since these supernovae are so bright they can also be observed in very distant galaxies, well outside the range of Cepheid variable estimation. From the apparent peak brightness of a particular type 1a supernova, knowing the actual peak luminosity then gives the distance. In this way distances of remote galaxies can be estimated out to three billion light years — a long way but still not the limit of the observable Universe.

24.3 The Expanding Universe

A very prominent worker in the field of determining the distances of galaxies, using Cepheid variables, was the American astronomer, Edwin Hubble (Figure 24.4). In 1929 he combined his distance results with some previous results found by another American astronomer, Vesto Melvin Slipher (1875–1969), on the velocities of distant galaxies, found by using the Doppler effect. Hubble's conclusion, based on the rather imprecise measurements available at that time, was that distant galaxies were receding from our own galaxy at a speed that was proportional to their distance. This conclusion, now known as *Hubble's law*, was reinforced by more precise later observations; a plot of velocity against distance based on distances derived from type 1a supernovae and high quality velocity estimates is shown in Figure 24.5.

Hubble's law had a profound effect on the way the Universe was perceived. It had been generally believed that, although changes were

How Matter Began

Figure 24.4 Edwin Hubble (1889–1953).

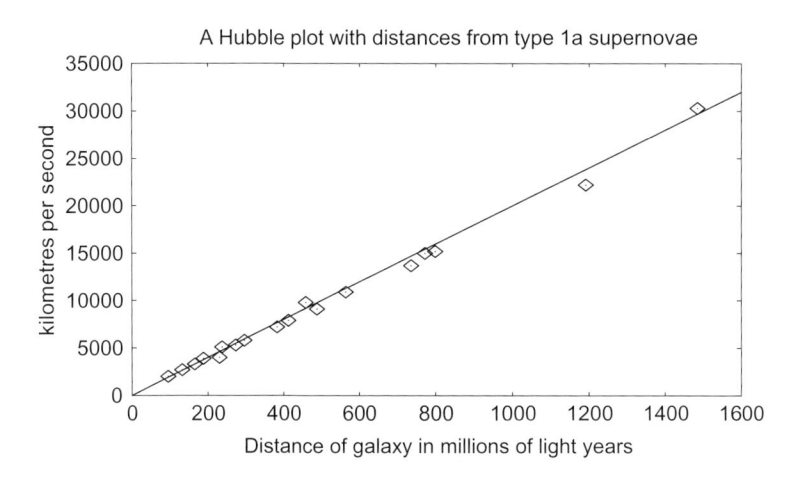

Figure 24.5 An illustration of Hubble's law.

going on within the Universe, on the large scale it was unchanging — at least in terms of extent. Now there was clear evidence that the Universe was expanding. There were galaxies for which no type 1a supernovae were observed but for which recession speeds could be measured and from Hubble's law their distances could be estimated. The largest red shift observed corresponds to a recession speed 0.93 times that of light. Since a speed greater than that of light is unobservable according to current theory, it seems that the observed Universe is approaching some theoretical limit.

24.4 The Big Bang Hypothesis

From the Hubble plot it is found that a galaxy at a distance of 3000 million light years is receding from our galaxy with a speed of 64 000 kilometres per second. The fact that the galaxy is getting further away with time does not necessarily mean that its line of motion passes through our galaxy. However, if we make the assumption that it *does* pass through our galaxy then, by imagining that the motion of the galaxy was reversed, we can calculate how long ago the two galaxies coincided. The calculation is done by dividing the distance by the speed, and it results in the answer of 14 billion years. No matter which galaxy we take the answer is always the same; at twice the distance there is twice the speed so dividing one by the other always gives the same result.

In 1931 the Belgian astronomer and Roman Catholic priest, Monsignor Georges Lemaitre (Figure 24.6), put forward a theory for the origin of the Universe, which he called the *Primeval Atom Theory*

Figure 24.6   Georges Lemaitre (1894–1966).

but later became known as the *Big Bang Theory*. This idea, which is suggested by Hubble's law, proposes that 14 billion years ago the energy of the Universe was suddenly created, with all the energy required to produce it concentrated into what was essentially a point, a point with no volume referred to by scientists as a *singularity*. From that point the Universe expanded outwards and 14 billion years later has evolved into what we observe today. The implication of this model is that, at the instant the Universe began, space and time did not exist. One cannot refer to a time *before* the Big Bang because there was no such thing as time before the Universe began. Again, one cannot ask what the Universe expanded into because the only space that exists is that *within* the bounds of the expanding Universe. These are very challenging ideas, outside the range of human comprehension, but the model is capable of being expressed in mathematical terms.

The Big Bang theory *is* just a model and is not accepted by all astronomers — although the vast majority of them do accept it. It has already been pointed out that it is not necessarily true that if the path of a galaxy was reversed it would collide with our own galaxy but what does seem clear is that, at the very least, some 14 billion years ago the Universe was far more compact and crowded than it is now.

On the basis of the Big Bang theory what was the Universe like at the instant it began? It was an unimaginable concentration of just pure energy — no matter could exist. Then it began to expand and, once this happened, then time and space came into existence. We shall now consider the processes that happened that gave us what we have now, given this postulated initial starting point.

24.5 The Creation of Matter

One of the primary goals of high-energy particle physics is to try to simulate conditions in the early Universe. However, there is simply not enough energy on Earth, even transmuting all the mass available into energy, to reproduce conditions at the very beginning of the process. So for the early stages of the model of the evolution of the Universe we must rely on guesswork and intuition. The various stages, according to the currently accepted model, are as follows.

From the Beginning to 10^{-12} Seconds

At this stage in the expansion of the Universe the laws that governed the behaviour of energy and matter are unknown and different from those that operate now. There would be no clear distinction between energy and matter and the forces that operated would be of an unfamiliar kind. During this period there occurred a rapid expansion of the Universe when the motion of the boundary was at a speed greater than that of light.

By the end of this period the Universe was densely filled with radiation and with particles and antiparticles of various kinds and the physical laws that exist today were beginning to operate.

From 10^{-12} to 10^{-10} Seconds

At the beginning of this period there was very little matter in existence. It was pointed out, and illustrated in Figure 16.2, that under the right circumstances an energetic photon can be transformed into an electron-positron pair. With *extremely* high-energy photons available it is possible to make other particle-antiparticle pairs and at this time quark-antiquark pair production could take place. At the same time as quarks and antiquarks were being made so they were being annihilated when they happened to come together.

As the Universe expanded the density of radiation became less and the temperature fell. With fewer very high-energy photons available the rate of quark-antiquark production reduced. Now something very strange happened for which there is no convincing explanation. The net result of quark-antiquark production and annihilation, and perhaps other unknown processes, was that the Universe was left with a greater number of quarks than antiquarks. That this was so is indicated by the fact that we live in a Universe consisting of protons and neutrons made from quarks and not one consisting of antimatter made from antiquarks. The excess of quarks over antiquarks was small but sufficient to give us all the material in the Universe. Another outlandish possibility, as an alternative to having unequal numbers of quarks and antiquarks, is that some separation occurred so that one part of the Universe had a slight excess of quarks (the part we live in) while another had a slight excess of antiquarks. In that case there would be parts of the Universe composed of antimatter. If that were so then we must hope that our galaxy stays well clear of those parts!

At this stage of the Universe's development quarks and antiquarks were giving rise to the various exotic baryons and mesons that can now be produced in high-energy physics collision experiments.

Time about 10^{-4} Seconds

Now that quark-antiquark production had ceased the quarks combined together in pairs to give various kinds of meson or in threes to give baryons including neutrons and protons. However, although the basic raw material to produce different kinds of atom was available it was still far too hot for protons and neutrons to come together to form atomic nuclei.

Time about 1 Second

Isolated neutrons are unstable and after some time, averaging about 15 minutes, a neutron will decay into three particles — a proton, an electron and an electron antineutrino. By contrast, protons are stable and in isolation will continue to exist indefinitely. However, it is possible to break up a proton by collision with another particle. One way is by the impact of an antineutrino that can give rise to a neutron and a positron. Another way is to bombard it with an electron, which produces a neutron and a neutrino. The ways in which neutrons and protons break down are illustrated in Figure 24.7.

To produce breakdown of a proton requires the energies of the colliding particles — antineutrinos or electrons — to be extremely high and this will only be so when the prevailing temperature is very high; the temperature of a substance is a measure of the average energy of the particles that comprise it. However, the temperature of the Universe was steadily falling and so the rate at which protons were being struck by particles and being converted into neutrons also fell. By contrast the disintegration of neutrons to give protons went on

284   *Materials, Matter and Particles*

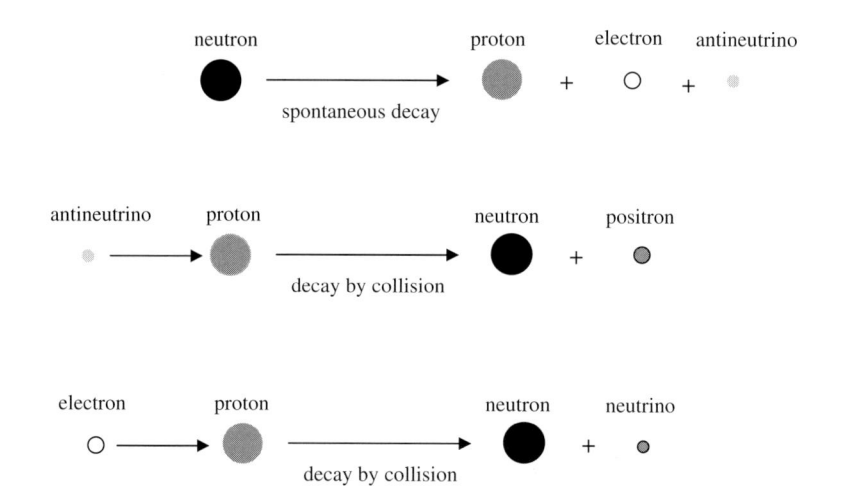

Figure 24.7 The spontaneous breakdown of an isolated neutron and the breakdown of a proton by two processes depending on collisions.

unabated — it was a spontaneous process. For this reason, during this period the ratio of protons to neutrons continuously increased.

Time about 100 Seconds

Although neutrons and protons had occasionally formed temporary associations prior to this time, the temperature was so high that they soon disintegrated. Another factor associated with high temperature is that the protons and neutrons were moving at such high speeds that they were close together for too little time for the bonding process to take place. By now, however, the temperature was low enough for protons and neutrons to come together to form the nuclei of light elements — helium (He) and lithium (Li) and the hydrogen isotope deuterium (D). These nuclei, previously shown in Figures 13.1 and 15.1 are shown again in Figure 24.8. At this stage there were about seven times more protons than neutrons in the Universe and the amounts of the light nuclei that were formed was governed by the availability of neutrons. Any neutrons that were absorbed into nuclei acquired stability and henceforth were permanent members of the Universe. Any that did not become part of nuclei, decayed and added to the number of protons.

deuterium   helium   lithium

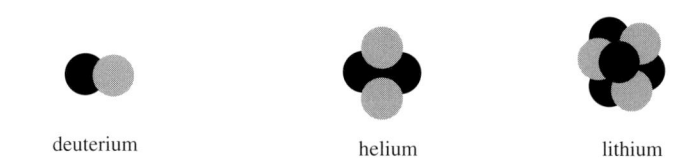

Figure 24.8 Light nuclei formed early in the expansion of the Universe (grey = proton, black = neutron).

Time 10 000 years

The story of the Universe thus far is that energy has been transformed into matter and by this time there was more energy in the form of matter than in radiation. In addition the Universe had cooled to such an extent that the available photons no longer had enough energy to be transformed to even the lightest particles. From now on matter and energy were independent of each other and as the Universe expanded so the mean density of matter became less as did the mean density of radiation. The Universe thus steadily cooled.

Time 500 000 years

Light-atom nuclei had been formed and large numbers of free protons existed but the attachment of electrons to these to produce complete atoms had not taken place. At the high temperatures that had previously existed the energy of the electrons, which also existed in large numbers, was far too high for them to attach themselves to nuclei. By this time the temperature had fallen to the point where electrons could attach themselves to nuclei to produce complete atoms. It will be recalled that there was a large surplus of protons in the Universe after all the available neutrons had been used to make nuclei of light atoms. With electrons attached these gave rise to the hydrogen, which was, and still is, the preponderant constituent of the Universe.

Starting with a vast amount of energy at a singularity, with no space and no time, we have progressed to a point where both time and space exist and where the Universe contains plenty of material. It is not a Universe that we would recognize since the material it contains is all in a dispersed form and with a mean temperature of 60 000 K compared with 3.2 K at present. However, the light elements that then existed were not the kind of material required to form the Earth and the other solid planets. To make these bodies, and what they contain, including us, other elements were needed, including oxygen, carbon, silicon and iron, all very common elements. We must now find out how these came into being.

Chapter 25

Making Heavier Elements

25.1 The Architecture of the Universe

When the night sky is explored with a powerful telescope many different kinds of object are observed. There are planets, satellites and comets but these are trivial local features and have little to tell us about the way the Universe evolved on the larger scale. Next in increasing size we see individual stars, sometimes single stars, but just as often in pairs to form a binary association in which the two stars orbit around each other. These individual stars or pairs of stars, moving in isolation in space, are known as *field* stars. There are also larger associations, typically containing between 100 and 1 000 stars, known as *galactic*, or *open*, *clusters*. A galactic cluster, the Pleiades, is shown in Figure 25.1. It contains seven very bright stars, so that the cluster is sometimes called The Seven Sisters.

The individual stars in galactic clusters can be seen because the stars are well separated in the field of view. There is another kind of stellar cluster containing many more stars, numbering in the range 100 000 to 1 000 000, where many of the stars, especially those towards the centre of the cluster, cannot be resolved. This kind of cluster is known as a *globular cluster* and a typical one, with the rather prosaic name M13, is shown in Figure 25.2.

On an even larger scale there are huge stellar systems known as galaxies containing of the order 10^{10} to 10^{11} stars. Galaxies come in a variety of shapes and sizes, the main types being *elliptical galaxies* and *spiral galaxies*. Elliptical galaxies, as their name suggests, are

Figure 25.1 The Pleiades cluster (John Lanoue).

Figure 25.2 The globular cluster M13 (ESA/NASA).

featureless blobs of a roughly elliptical shape. The Sun is situated about one-third of the way out, some 25 000 light years, from the centre of our home galaxy, the *Milky Way*, a quite large spiral galaxy containing about 2×10^{11} stars. Since the Earth is embedded within the Milky Way we can only infer its shape indirectly but there are many similar spiral galaxies that we can see clearly and that give an idea of its appearance. The main structure of a spiral galaxy is in a fairly thin disk-like form. Figure 25.3 shows the Pinwheel Galaxy, a very handsome spiral galaxy seen perpendicular to the disk. Figure 25.4

Figure 25.3 The Pinwheel Galaxy (ESA/NASA).

Figure 25.4 The Sombrero Galaxy (HST/ESA/NASA).

shows the Sombrero Galaxy, a spiral galaxy seen edge on. It will be seen that in the centre of the disk there is a bright bulge, called the *nucleus* and that there is also a diffuse flattened sphere of stars and stellar clusters that encompasses most of the disk, known as the *halo*.

Galaxies are, on the whole, very well separated from each other but there are occasional close approaches, and even collisions, of pairs of galaxies. However, when the Universe is examined on a very large scale it is found that there are *clusters of galaxies* distinctly separated from other clusters of galaxies. The Milky Way is a member of a cluster of 35 galaxies, known as the *Local Group*, which has a diameter of 10 million light years, 100 times that of the Milky Way itself. It is claimed that on an even larger scale there are clusters of clusters of galaxies well separated from other clusters of clusters and perhaps this hierarchical structure of the Universe extends even further!

25.2 Dark Matter and Dark Energy

The Universe contains a great deal of substance that we cannot see. Galaxies are rotating systems; in the case of the Milky Way the period of one complete rotation is 200 million years but it is such a large system that, despite its slow rotation, stars at its periphery are moving at 500 kilometres per second relative to the galaxy centre. We know the laws of orbital motion, first described by Isaac Newton. The Earth moves round the Sun in a more-or-less circular orbit and it is the gravitational attraction of the Sun that prevents it from flying off into space. However, when we estimate the mass of the Milky Way from the material that we can see in the form of stars and stellar clusters it seems that there is far too little mass present to prevent the galaxy from flying apart. There is a similar story when it comes to clusters of galaxies. These are stable systems; otherwise we must make the unlikely assumption that we are observing them at a period in the history of the Universe when they just happen to exist in a clustered configuration. The galaxies in a group move relative to each other and the estimated masses of the galaxies in a group are insufficient to stop the group from flying apart. The conclusion from these observations is that the mass we can see accounts for only about 5% of the total mass of the Universe. The rest has been called *dark matter* and what it consists of is a subject of much debate and investigation. Some of it may be in the form of dead stars that have cooled to the point of being unobservable except through their gravitational influence. Another possibility is that they are objects of less than stellar mass that are too cool to be detected visually or by radiation-detecting instruments. A third possibility, under active investigation, is that it is due to the existence of rather exotic particles called Weakly Interacting Massive Particles (WIMPs). These are heavy, by particle standards, but only weakly interact with ordinary matter so thus far they have not been detected — although they show their existence by their gravitational effects.

Recent observations of the expansion of the Universe have suggested that the rate of expansion may be *increasing* with time. This is a very counterintuitive conclusion since the expectation is that gravitational forces pulling matter inwards should be *decreasing* the rate of expansion of the Universe. It has been suggested that, in addition to dark matter, the Universe also contains *dark energy* the pressure of which is overwhelming gravity and pushing matter outwards.

25.3 The Formation of a Star

Stage 1 The Initial Collapse

The basic unit from which all the major structures in the Universe are formed is the star. The formation of a star begins with a large diffuse gas cloud that is separated from neighbouring material and is collapsing under the influence of gravity. At this stage of the formation of a star, when it is still not a luminous object, it is customary to refer to it as a *protostar*. A typical starting point for a solar-mass star would be a cloud of radius 3.6×10^{11} kilometres and a density of 10^{-14} kilograms per cubic metre. To give an idea of what this means the radius of the cloud is 2 400 times the distance of the Earth from the Sun, a unit of distance that astronomers call the *astronomical unit*, while the density is roughly equivalent to the density of the gas in the best ultra-high vacuum that can be achieved in scientific laboratories. However, by comparison with the density of the matter between stars, the *interstellar medium*, which has a density of 10^{-21} kilograms per cubic metre, the cloud has a significantly high density.

The cloud begins to collapse, very slowly at first, but as it collapses the gravitational force pulling inward becomes stronger so that the rate of collapse increases with time. The effect of the collapse is to compress the material of the cloud into a smaller volume and the compression of a gas generates heat. This is a well-known phenomenon to anyone who has pumped up a bicycle tyre; the air in a bicycle pump is compressed until its pressure can drive it through the valve into the inner tube. The repeated compression at each stroke of the pump can make the barrel of the pump quite hot. In the case of the collapsing cloud, although heat is generated within it, its temperature does not rise at first because the cloud is so transparent that it radiates the heat away and it stays at the same temperature as its surroundings. In most locations where stars are being born this would correspond to a temperature between 20 and 50 K. This state of temperature constancy lasts until the protostar has reached about 2% of its original radius. The development of the protostar during this stage of its collapse is marked as the full-line in Figure 25.5, which shows the variation of temperature with radius. Its total duration is about 20 000 years.

Stage 2 Heating Up

By the time the star reaches the end of the full-line it is collapsing quickly and because its density is increasing it is getting ever more opaque to the passage of radiation. For this reason the heat developed by the collapse is increasingly retained, the cloud heats up and the resulting increase in pressure slows down the collapse and brings it

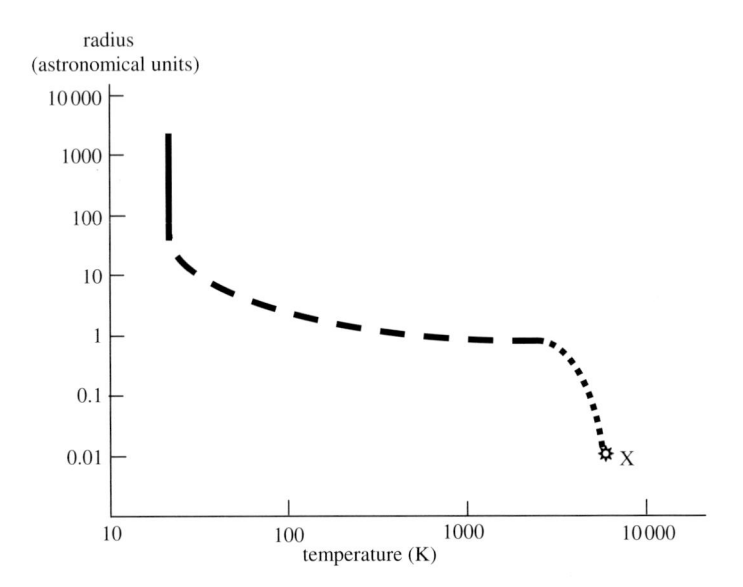

Figure 25.5 The variation of radius and temperature in the passage from a protostar to a main-sequence star.

almost to a halt. By the end of this stage of collapse the radius of the protostar is slightly less than one astronomical unit and its temperature has reached about 3 000 K. This part of the collapse is shown as the dashed-line in Figure 25.5. At the end of this stage, which lasts only a few hundred tears, we can stop using the term protostar and use *young star* instead.

Stage 3 Slow Collapse to the Main Sequence

The young star is now in a state of near equilibrium where gravity, which tends to make the young star collapse, is being balanced by the pressure, due to its temperature, that tends to make it expand. However, because the surface of the young star is hot it is radiating energy out into space. Instinctively we might think that since it is losing energy by radiation it should cool down but, paradoxically, it does the exact opposite and heats up. This is because it slowly collapses and the gravitational energy released by the collapse provides both the energy to radiate away and also that to heat up the young star.

The temperature is not uniform throughout the young star since it is much cooler on the outside than internally and the temperatures indicated in Figure 25.5 are the surface temperatures as seen by an outside observer. The dotted line in Figure 25.5 indicates the progress of the young star during this period. When it reaches the end of the dotted line, at point X, the star enters a new stage in its development, which is one of very long duration. For a solar-mass star the time for the slow contraction stage is 50 million years. A star of 2 solar masses will spend only eight million years in this contracting stage while a star of one-half a solar mass will take 150 million years. It is a general rule that the more massive a star the faster it moves through the various stages of its development.

Stage 4 The Main Sequence

At the centre the internal temperature has risen to 15 million K, at which stage nuclear processes take place in the interior that transform hydrogen to helium. The steps in this process were described in §17.9. Now there is a new equilibrium set up. The surface temperature of the star is such that it radiates energy away at the same rate as it is being internally generated and the star remains at the same radius. At this time in a star's development it is at the beginning of a state known as the *main sequence*. The Sun has been on the main sequence for about 4.5 billion years and will continue to be a main sequence star for another 5 billion years. It will then run out of hydrogen in its central regions and will enter a new stage of development. Its surface temperature is about 5 800K, which gives it a yellowish colour. More massive main sequence stars have a higher surface temperature and may seem white, or even bluish when they are very hot, while lesser mass stars are cooler and seem to have a reddish hue.

25.4 The Death of a Star

Stage 1 Hydrogen Shell Burning

Since new stars consist mainly of hydrogen they have an abundance of fuel to maintain the main-sequence stage for a considerable time. The hydrogen is being converted to helium in the innermost part of the star and eventually hydrogen will become depleted there, its place being occupied by helium. Since energy production in the centre of the star is reduced through lack of hydrogen there is less pressure generated to resist gravitation forces so the core starts to contract. The contraction generates heat until the temperature reaches the level at which material around the central core, still hydrogen rich, is able to sustain nuclear reactions. This gives nuclear reactions in a shell that slowly moves outwards as the inner region depleted of hydrogen becomes larger. The star is now in the state of *hydrogen-shell burning*, which is illustrated in Figure 25.6. Also shown in the figure is the pressure force exerted by the shell burning on neighbouring material. It acts inwards on the core material compressing it to a higher density and outwards on material further out causing the star to expand. The star becomes a red giant, the kind of star that, in association with a white dwarf, gives rise to a type 1a supernova, as described in §24.2.

Making Heavier Elements 295

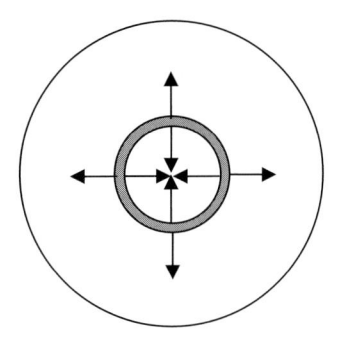

Figure 25.6 Hydrogen-shell burning and the generated forces.

Stage 2 Forming a Carbon Core

While hydrogen-shell burning is progressing the core is virtually all helium, which is being compressed by ever-increasing pressure and hence is steadily increasing in temperature. When the core reaches a temperature of approximately 100 million K the helium in the core begins to take part in nuclear reactions, the effective end result of which is that three helium nuclei fuse together to become a carbon nucleus — the two-stage *triple-alpha reaction* described in §17.4 and illustrated in Figure 25.7.

With the onset of the generation of energy in the core by the triple-alpha reaction the situation returns to something like the initial main-sequence stage where energy was generated at the centre of the star. Consequently the star collapses again to a more compact state, similar to, but not the same as, the main sequence. Now a similar sequence of events happens where helium is depleted in the core, which is mostly carbon, and helium-shell burning takes place with the addition of a hydrogen-burning shell further out. Helium burning gives far more energy than hydrogen burning so now, not only does the star expand into a new red-giant form, but it also sheds some of its outer material. Figure 25.8 shows a star in this state ejecting material in a form known as a *planetary nebula* — which has nothing to do with planets.

296 *Materials, Matter and Particles*

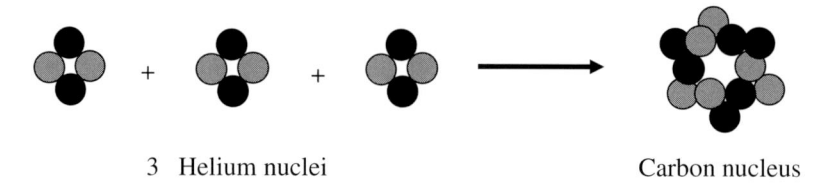

3 Helium nuclei → Carbon nucleus

Figure 25.7 The triple-alpha reaction. grey = proton, black = neutron.

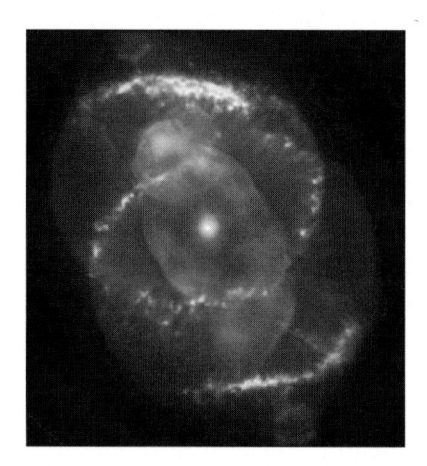

Figure 25.8 The Cats-eye Nebula (HST/ESA/NASA).

Stage 3 The End Game

What next happens to the star depends on its mass. For a star of solar mass or less all the outer material is shed and the very dense core is left behind in the form of a white dwarf, which mostly consists of carbon. The material in a white dwarf has a very high density, the state of which is known as *degenerate matter*. In this state the electrons form a system governed by the laws of quantum mechanics (Chapter 12) and the Heisenberg uncertainty principle (§12.4); the atomic nuclei reside independently within this system.

For higher mass stars the temperature within the core becomes so high that other nuclear reactions can take place involving heavier elements. Some such reactions are:

        helium + carbon ⇒ oxygen
        carbon + carbon ⇒ neon + helium
        carbon + carbon ⇒ sodium + proton
        carbon + carbon ⇒ magnesium + neutron
        oxygen + oxygen ⇒ silicon + helium.

There are many reactions generating a succession of heavier elements and a massive star may go through several stages of expansion and contraction as new episodes of shell burning occur and new reactions take place at the centre. At some stages the material stripped from the outside may leave behind a white dwarf, but now one that will contain elements heavier than carbon. This process of building up heavier and heavier elements stops at the stage when iron is produced. The reason for this is that in all the nuclear processes that occur up to the formation of iron the masses of the final products are less than the initial mass and the difference of mass appears as an output of energy. This heats up the stellar material to give rise to the possibility of new reactions. However, for reactions involving iron or heavier elements the final products have a greater mass than the reacting components and hence, to provide the extra mass, an *input* of energy is required. This is logically impossible; if energy were taken out of the system to give such reactions then the system would cool down and prevent such reactions. Figure 25.9 shows a highly evolved massive star at the stage when it has developed an iron core. Elements other than those shown, up to iron, will have been produced.

If the star is very massive, greater in mass than the Chandrasekhar limit mentioned in §24.2, i.e. 1.4 times the solar mass, then the pressure in the iron core becomes so great that the iron nuclei cannot retain their normal structure. The protons combine with free electrons to become neutrons, the neutrons take on the dense form of degenerate matter, and the core suddenly contracts to a diameter

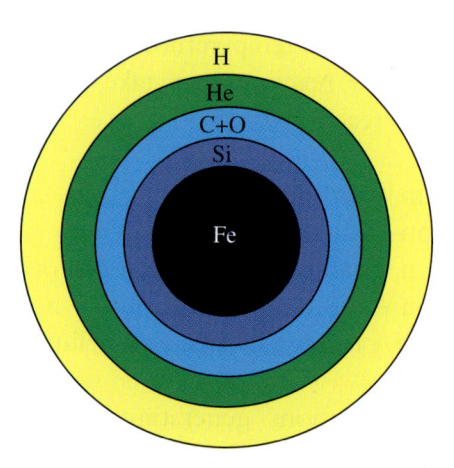

Figure 25.9 Composition shells in a highly evolved massive star (not to scale). H = hydrogen; He = helium; C = carbon; O = oxygen; Si = silicon; Fe = iron.

of a few kilometres. Outer material rushes in at high speed to fill the vacated space and then violently impinges on the newly formed neutron core. The result is a violent explosion, a supernova, which throws all the material outside the core into space. What is left behind is a *neutron star*, an object with stellar mass and a diameter of a few kilometres. It consists of a sphere of tightly packed neutrons covered by a thin iron crust. The density of a neutron star is such that a teaspoonful of it has a mass of 5 000 million tonnes.

A somewhat more massive star, of about three solar masses or more, would leave behind an even more exotic residue. The total mass of the neutron core would be so great that even the neutrons would be crushed by gravity. All the mass would collapse to form a *black hole*. This is an object of finite mass but, theoretically at any rate, of zero size. No radiation can escape from it, so it cannot be seen, but it can be detected through its gravitational field.

The supernova phenomenon enables the process of element formation to break through the iron barrier. A supernova is an enormous source of energy and this available energy enables reactions to take place involving iron and heavier elements. A combination of the processing within a star and the effect of a supernova enables the formation of all the isotopes of elements up to uranium-238 and beyond; the unstable ones with shorter half-lives will eventually disappear. The supernova explosion ejects all this material into space so that it becomes a component of the galactic gas available to be incorporated into other stars.

25.5 The Evolving Universe

In Chapter 24 we traced the development of the Universe to an age of 500 000 years at which time it was full of hydrogen, deuterium, helium and lithium dispersed throughout its volume. We have also seen how, once stars are formed, this primeval material can be processed to give all the elements that now exist. Now we consider the way that the primeval material could collect together in such a way that stars could form.

After 500 000 years the Universe had reached a considerable size, even if small by comparison with its present size. Turbulence would have developed in the expanding material so that separate streams of the gaseous material would have been colliding from time to time. When gas streams collide they compress each other and form a higher density region and in this way the Universe would have developed a lumpy structure. Some of the lumps may have been as massive as clusters of clusters of galaxies and the effect of gravity would be to concentrate each of them and to separate it further from its neighbours. Within each of the condensations further turbulence would arise and a hierarchy of lumpy substructures would have developed. Eventually there were condensations with the masses of galaxies forming; the hierarchy of condensations — collapsing clusters of clusters, clusters of galaxies and individual galaxies are illustrated in Figure 25.10.

By the time the galaxy-mass condensations had formed — the dots in Figure 25.10 — a new mechanism would come into play, one that was first described by James Jeans (1877–1946). He considered the behaviour of a spherical mass of gas at a particular density and temperature. The pressure exerted by the gas, that is tending to disperse the material, depends only on the density and temperature but the gravitational force tending to keep it together increases with the

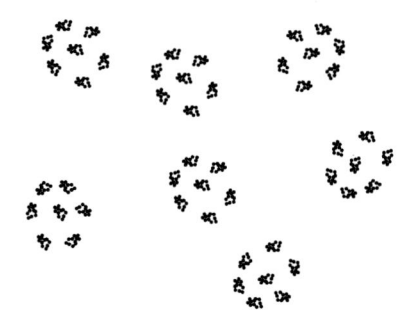

Figure 25.10 A schematic representation of galaxy-mass condensation (small dots), clusters of galaxy-mass condensations and clusters of those clusters.

size of the sphere, and hence its mass. For any particular combination of density and temperature there is a *Jeans critical mass* corresponding to the minimum mass that will preserve the integrity of the mass of gas and prevent it from expanding and dissipating.

There is another scenario in which the Jeans critical mass comes into play. If there is a large volume of gas at uniform density and temperature then it will be intrinsically unstable and will spontaneously break up into separate blobs each with the Jeans critical mass. Calculations indicate that the conditions within an original galaxy condensation are such that the Jeans critical mass is equivalent to about half a million solar masses, or about the mass of a globular cluster. When the globular-cluster mass collapses and the material begins to move at greater than the speed of sound in the gas then turbulence sets in and the collision of turbulent streams of gas can produce local concentrations of gas that can collapse further under gravity to form stars, as described in the previous section.

The low density in the early stages in the collapse of a globular-cluster cloud gives a high Jeans critical mass so that the first stars produced would have been massive. Such massive stars will go through their evolution very quickly — for a 50 solar mass star less than one million years — after which time they will explode leaving behind a black hole and throwing out material containing elements covering the full range of the Periodic Table and even more massive unstable nuclei beyond the Periodic Table. These postulated early stars, consisting only of the material produced in the Big Bang — hydrogen, deuterium, helium and lithium — are referred to as Population III stars. From the Fraunhofer lines in a stellar spectrum it is possible to find not only what elements are present in a stellar atmosphere but also the amounts of each element. The degree to which a star contains heavier elements is referred to as its *metallicity*, although many of the elements contributing to this property are not metals. Astronomers are always on the lookout for Population III stars that would have zero metallicity but if, as it is assumed, they were very massive and quickly evolved, then such searches are unlikely to succeed.

The later stars formed in a globular cluster will contain some of the heavier elements produced by the extinct Population III stars and the later they are produced the more heavy elements they will contain. Globular cluster stars, all of which contain only slight amounts of heavier elements, up to about 0.2% by mass, are referred to as Population II stars. An interesting observation from globular clusters is that they contain no main-sequence stars with masses greater than about 0.8 times the mass of the Sun. The implication of this is clear. Globular clusters originally contained stars with a wide range of masses, including many with masses well above the present 0.8 solar-mass observational limit, and those with mass greater than that limit have completed their evolution, leaving behind a black hole, a neutron star or a white dwarf. The time for the complete evolution of a 0.8 solar mass star is almost the same as the age of the Universe showing that globular clusters formed shortly after the Universe came into being.

Stars move around within clusters changing their direction, accelerating and decelerating as they gravitationally interact with each other. Once in a while a star near the boundary of a cluster will, through the interaction of other stars, reach a speed greater than the escape speed from the cluster and will leave it. When a theoretical calculation is made of how long it would take for a globular cluster to completely dissipate by this process the answer comes to many times the age of the Universe. We can be reasonably certain that all globular clusters formed early in the life of the Universe still exist today — although they may have lost a small number of their original members. When the same theoretical calculation is made for galactic clusters it is found that their lifetimes are about one hundred million years, of order 1% of the age of the Universe. It is therefore clear that all the galactic clusters observed are of recent origin, and that galactic clusters have been forming and dispersing over most, if not all, of the time that the Universe has existed. While globular clusters may occur in any location in a galaxy — in the core, the disk or the halo — galactic clusters are only found in the disk. Galactic clusters must be constantly forming and dispersing in the disk. Indeed, the process of star formation may be observed at the present time in regions where there are dusty dense cool clouds. A typical star-forming region is the Orion Nebula, a dense dusty cloud south of Orion's belt; Figure 25.11 shows the Trapezium Cluster within the Orion Nebula, a very active region of star formation.

Stars in a galactic cluster, and field stars produced in the same way, will have had their material processed several times within previous stars that have gone through their evolutionary paths. They are referred to as Population I stars and their metallicities are much

Figure 25.11   The Trapezium Cluster.

higher than those of Population II stars. The Sun, a typical Population I star, has about 2% of its mass in the form of heavier elements.

In recent years many planets have been found around nearby main-sequence stars, all of the Population I variety, and it is generally accepted that the material of these planets will have come from the same source as that which formed the parent star. This is how the material that formed our world came into being — its atmosphere and solid substance, the biosphere and everything within it, including us. An eminent American astronomer, Carl Sagan (1934–96), a great expositor of popular science, once described humankind as being evolved from star stuff. How right he was!

Look around you — look in a mirror. All the matter that you see, living and non-living, has been through the violence of several supernovae, and may do so many times more in the distant future.

Index

α-particle emission   116
α-particle induced reactions   155
α-particle scattering   74, 75
α-particles   72, 73, 89, 141
β⁺-particle   148
β-decay   143, 148
β-emission   143
β-particles   72
γ-ray induced reactions   157
γ-ray photon   144
γ-rays   72

adaptation   184
adenine   199
Ahnighito   223
air   34
Akasha   15
alchemy   18, 23
alleles   208
alum industry   49
aluminium   139
aluminium-26   157
amino acid   201
ampere   55
amplitude   109

Anaximenes   11
Anderson, Carl D.   145
Ångstrom unit   195
angular momentum   95, 115, 149
aniline   50
animal intelligence   214
animalia   193
Anne Boleyn   209
anode   46
antineutrino   150, 154
antineutron   267
antiparticle   144, 146, 267
antiproton   267
antiquarks   269
archaea   190
argon   124
arsenical bronze   220
astatine   139
astronomical unit   291
atom   13
atomic mass units (amu)   94, 147
atomic number   85, 87
autosomes   207
Avogadro, Amedeo   45
Avogadro's law   45

Avogadro's number   47
azimuthal quantum number   104, 116

bacteria   187
Balmer series   98
Balmer, Johann   98
barium platinocyanide   66
barium-137   143
Barkla, Charles   82
baryon   283
basal body   189
Becquerel, Henri   68
bending magnets   264
beryllium   91, 124
Berzelius, Jöns Jacob   46
Big Bang Theory   281
binding energy   147
bismuth   139
black hole   298
blue shift   275
Bohr, Neils   95
Boltzmann, Ludwig   47
boron   124, 164
Bose, Satyendra Nath   134
Bose-Einstein statistics   134
boson   134, 258
Bothe, Walter   91
Bouguereau, William-Adolphe   231
Boyle, Robert   25, 53
Boyle's law   28
Bragg reflection   180
Bragg, William Henry   180
Bragg, William Lawrence (Lawrence)   178, 181

Bristlecone pine   184
Bronze Age   215, 221
bronze   16
Brooklyn National Laboratory   265
Buckminsterfullerene   251

cadmium   164
cadmium-109   152
calcite   180
calcium   124
calcium-137   143
calx   31
canal rays (kanalstrahlen)   88
canopic jars   17
capsule   187
carbon dating   153
carbon dioxide   34
carbon fibre   243
carbon isotopes   138
carbon nanotubes   175, 252
carbon   124, 135
carbon-14   152
Carothers, Wallace   240
cast iron   225
cathode rays   55, 58
cathode   46
cell wall   188
cellophane   239
cellulose   238, 240
cement   227
centrifuge   165
Cepheid variable   276
CERN   271
Chadwick, James   92
Chaka   225

Chandrasekhar limit   278, 297
characteristic X-radiation   83, 86
charcoal   31, 33
Charles, Jacques   43
Charles' law   43
chemical bond   48
chemical elements   29, 43
Chernobyl   166
Chicago Pile-1   164, 165
chimpanzee   193
chlorophyll   192
chromosome   199, 207
classical mechanics   111
clevite   73
clusters of galaxies   289, 299
Cockroft, John   259
Cockroft-Walton generator   259
codon   201
compound microscope   172
compounds   43
concrete   227
conductivity and temperature
   conductors   246
   semiconductors   246
conductivity   34
consciousness   194
conservation rules   154
control rods   164
Cookes dark space   54
copper   219
coral   183
corpuscle   61
cosmic rays   146, 153, 258
cotton   234

coulomb   55
covalent bonding   127
Cowan, Clyde I.   150
Cowan-Reines experiment   151
Crick, Francis   203
Cromwell, Oliver   26
Crookes tube   56
Crookes, William   56
Curie, Marie (Maria Sklodowska)   70
Curie, Pierre   70
cyclotron   261
cytoplasm   188
cytosine   199

dalton   44
Dalton, John   42
dark energy   291
dark matter   290
Darwinian evolution   185, 203
Darwinian selection   210
Davisson, Clinton   100
Davisson-Germer experiment   100
Davy, Humphrey   54
de Broglie hypothesis   100, 113
de Broglie wavelength   108
de Broglie waves   102
de Broglie, Louis   99, 101, 175
degenerate matter   296, 297
Democritus   13, 43
deoxyribonucleic acid (DNA)   188
dephlogisticated air   35
deuterium   136, 167, 284
D-glucose   238

diamine   240
diamond   243
dicarboxylic acid   240
diffraction   82
diffraction grating   177
diode   250
Dirac, Paul   144, 154
DNA (deoxyribonucleic acid)   188, 198
DNA model   205
Dog (meteorite)   223
dominant gene   208
Doppler shift   273
Doppler, Christian   273
double bond   48
double helix   203, 205
doubly-ionized atom   136
Drexler, Eric   253
Dumas, Jean Baptiste André   46
Dupont   240

e/m   60
$E = mc^2$ 142, 260
Earth heating   142
Einstein, Albert   81, 97, 99, 103, 163
electric charge   64
electrolysis   46, 54
electromagnetic radiation   77
electron antineutrino   268
electron density map   196
electron micrograph   175
electron microscope   174
electron neutrino   268
electron shells   121
electron volt   156
electron wave functions   113
electron   268
electronegative element   46, 126
electronic structure   125
electron-positron pair   145
electropositive element   46, 126
elements of nature   5
elements   39
elixir   19
elliptical galaxies   287
Ellis, Charles Drummond   148
Empedocles   12
energy levels   97
ether   25
eugenics   210
eukaryota   191
Ewald, Paul   177, 181
exchange particle   255
extrinsic semiconductor   248

Faraday dark space   54
faraday   55, 57
Faraday, Michael   53
Faraday's law   57
fast hydrogen atoms   89
fast nitrogen   90
Fermi, Enrico   134, 159
Fermi-Dirac statistics   134
fermion   134, 144, 149, 154, 258
Feynman, Richard   250
field star   287
fire   215
fission reactions   159
flagellum   189
float glass   218
fluorescence   56, 66, 68

fluorine   124
fluorite   174
fluorspar   180
four elements   12
Franklin, Benjamin   34
Franklin, Rosalind   203
Franklin-Wilkins Building   205
Fraunhofer lines   274
French revolution   33
Friedrich, Walter   178
Frisch, Otto   161
Frisch-Peierls memorandum   162
fruit fly   184
fullerenes   251
fundamental particles   268
fungi   191
fur   230
fusion power   168
fusion   167

galactic (open) cluster   287, 302
galaxies   299
Galen   24
Galileo Galilei   25, 173
Gay-Lussac, Joseph   43
Geber (Jabir ibn Hayyan)   18
Geiger counter   73, 93
Geiger, Hans   73
Geissler tube   56
Geissler, Heinrich   56
Gell-Mann, Murray   269
gene therapy   211
gene   199
genetically modified (GM) food   211
genome   199

genomics   202
Gerhardt, Charles Frédéric   47
Gerlach, Walther   129
Germer, Lester   100
glass   216
glassblowing   218
globular cluster   287, 301
gluon   271
Goldstein, Eugen   88
Goudsmit, Samuel   131
grape sugar   238
graphite   243
graviton   271
greenhouse gas   190
ground state   96
growth   185
guanine   199
gypsum   227

haematite   222
haemoglobin   202
Haga, Hermann   82
Hahn, Otto   148, 160
half-life   142
halo   289
Hanford reactor   164
Harkins, William Draper   91
harmonic oscillator   111
Heisenberg uncertainty principle   119, 296
Heisenberg, Werner   94, 118, 131
helium   73, 124, 284
helium-shell burning   295
Helmholtz, Herman Ludwig   57

Henry IV   20
Henry VIII   49, 209
herbal medicines   20
Hertz, Heinrich   58
Higgs particle   271
high-carbon steel   225
Hiroshima   165
Hittorf, Johann Wilhelm   56
hole   247
homo sapiens   193, 231
Hooke, Robert   173
Hubble, Edwin   278
Hubble's law   278
hydrogen atom, mass   61
hydrogen bomb   168
hydrogen   34, 124, 285
hydrogen-shell burning   294

IG Farben   50
Iijima, Sumio   252
India   225
interaction particle   271
interstellar medium   291
intrinsic semiconductor   247
Inuit   223, 230
Invisible College   26
in-vitro fertilization   211
ion   94
ionic bonding   126
Irish linen   233
Iron Age   222, 234
iron meteorites   222
iron pyrites   180
iron smelting   223
iron   219, 297
isotope   135

Jabir ibn Hayyan (Geber)   18
Jeans critical mass   300
Jeans, James   299
Juliot Curie, Frédéric   92
Juliot Curie, Irene   92
Jurassic Park   199
Justinian   234

Kanada   15
kanalstrahlen (canal rays)   88
kaons (K-mesons)   271
Kaalijärv   223
Kelvin (Lord)   47
kinetic energy   96
kingdoms   191
Knipping, Paul   178
Kroto, Harry   251

lambda particles   270
Large Hadron Collider (LHC)   266
laudanum   24
Laurent, Auguste   46, 47
Lavoisier, Antoine   36, 41
law of definite proportions   42, 44
Lawrence, Ernest   261
Leavitt, Henrietta   276
Lemaitre, Monsignor Georges   280
leptons   268
Leucippus   13, 43
life   183
light year   276
lime mortar   227
lime plaster   227

lime   227
LINAC (linear accelerator)   262
linear harmonic oscillation   109
linen   233
lipid   188
Lippershey, Hans   172
lithium   124, 284
lithium-6   168
Litvinenko, Alexander   143
Local Group   289
Lockyer, Norman   72
Loschmidt number   47
Loschmidt, Josef   47
luminosity   276
Lyman series   98
Lyman, Theodore   98
lysozyme   196

M13   287
Magdeburg hemispheres   27
magnesium isotopes   138
magnetic confinement   168
magnetic quantum number   105, 116
magnetite   222
magnifying glass   171
main sequence   277, 294
malachite   220
Manhatten Project   163
manufactured fibres   238
Marsden, Ernest   76
mass conservation   38
matrices   119
mauvine   50
Maxwell, James Clerk   103
Meitner, Lise   148, 160

Mendeleev, Dmitri   51
mercury, sulphur and salt   25
meson   266, 283
Mesopotamia   218, 219
messenger ribonucleic acid (mRNA)   188, 201
metabolism   186
metallicity   301
Meteorological Observations and Essays   43
Meyer, Julius Lothar   51
microscope   171
mild steel   225
Milky Way   288
Millikan, Robert   62
Millikan's oil drop experiment   63
minimum energy principle   121
moderator   162
molybdenum steel   227
momentum   99
monochromatic X-radiation   84
mordant   49
Morse code   198
Moseley relationship   87
Moseley, Henry   86, 180
mummification   17
muon neutrino   268
muon   258, 268
mutation   185, 210
mycoplasma   188

Nagasaki   165
nanotechnology   250
natron   17
Neanderthal man   231
neon isotopes   138

neon   124
neon-22   148
neutrino   149, 154
neutron decay   283
neutron induced reactions   158
neutron instability   143
neutron mass   94, 147
neutron star   298
neutron   91
New System of Chemical Philosophy   45
Newlands, John   51
Newton, Isaac   21, 27, 103, 255
Niger   225
nitrogen fixation   190
nitrogen   37, 124
n-type semiconductor   247
nuclear physics   89
nuclear power station   166
nuclear reactor   165
nucleon   136
nylon   240

obsidian   217
open (galactic) cluster   287
optic nerve   169
orbitals   115, 124
Orion Nebula   302
Ötzi the Iceman   219, 231
oxygen isotopes   138
oxygen   35, 124

PAN (polyacrylonitrile)   244
Paracelsus, Philippus Aureolus   24
parallax method   276
Parthenon   228

Pauli exclusion principle   133
Pauli spin matrices   131
Pauli, Wolfgang   131, 149
Peary, Robert   222
Peierls, Rudolf   162
Periodic Table   51, 53, 85, 88, 121, 123, 133, 300
Perkin, William   50
PET (polyethylene terephthalate)   242
phase   171, 195
philosopher's stone   19
phlogiston theory   30, 35
phosphate group   199
phosphorescence   68
photoelectric effect   80
photoelectron   81
photon   81, 97, 271
pili   189
Pinwheel Galaxy   288
pion (pi-meson)   258
pitchblende   70
Planck quantum of energy   110
Planck theory   108
Planck, Max   77, 99
Planck's constant   78, 95
planetary nebula   295
plantae   192
plasma membrane   188
plasma   58, 168
plastics   238, 240
Pleiades   287
Plucker, Julius   56
plum pudding model   61, 76
plutonium   164
Poincaré, Henri   68

polonium   71, 91
polonium-210   143
polyesters   242
polymer   199
polyps   183
polysaccharide   187
Pope John XXII   20
Population I stars   302
Population II stars   301
Population III stars   301
Portland cement   228
positron   144, 146
potassium uranyl sulphate   69
potassium   124
potential energy   96
Powell, Cecil   258, 268
Priestley, Joseph   32
Primeval Atom Theory   280
principal quantum number   104, 114
Principles of Chemistry   51
probability density   111
probability wave   110
promethium   139
protista   191
proton induced reactions   157
proton mass   147
proton   89
proton breakdown   283
protostar   291
Proust, Joseph   42, 44
Proust's law   42
p-type semiconductor   248
purine   199
pyrimidine   199
pyrophor   38

qi   12
quantum mechanics   111
quantum numbers   113, 121
quantum physics   80
quark flavours   269
quark-antiquark pair production   282
quarks   269
quartz   174
quintessence   12, 194
Q-value   156

radial probability density   114
radioactive carbon   153
radioactivity   70, 94
radiograph   67
radium   71
Ramses II   233
Ramsey, William   73
Rational Dissenter   33
rayon   238
recessive gene   208
rectifier   250
red algae   191
red giant   277, 294
red mercuric oxide   34
red shift   275
regeneration   185
Reign of Terror   39
Reines, Frederick   150
reinforced concrete   229
Relativistic Heavy Ion Collider (RHIC)   265
reproduction   184
residue   201
response to stimuli   186

retina   169
ribosomes   188
Richter, Jeremias   41
Robinson, Elihu   42
Romans   228
Röntgen, Wilhelm Conrad   65
Roosevelt, Theodore   163
Rosalind Franklin Award   205
Rosalind Franklin Building   205
Ross, Sir John   223
rotaxanes   253
Royal Institution   54
Royal Society of London   29
Rutherford, Ernest   71, 89, 95, 155, 259, 261
Rydberg constant   98

Sagan, Carl   303
scanning electron microscopy   175
Schrödinger wave equation   107, 109, 131
Schrödinger, Erwin   107, 144
Schweppe, Jean Jacob   34
selective breeding   210
semiconductors   245
short-range nuclear force   135
Si Ling Chi   233
silica glass   217
silica   217
Silicon Glen   246
Silicon Valley   246
Silk Road   234
silk   233
simple microscope   172
singly-ionized atom   136

singularity   281
slag   223
Slipher, Vesto Melvin   278
slow neutrons   159
soda cellulose   239
soda glass   217
soda-lime glass   217
sodium chloride   180
sodium   124, 139
sodium-22   148
Sombrero Galaxy   289
Sommerfeld orbit   124
Sommerfeld, Arnold   103
space   281
spagyrists   28
spin quantum number   132
spiral galaxies   287
stainless steel   227
Stanford linear accelerator   262
steel   20, 225
Stern, Otto   129
Stern-Gerlach experiment   130
stoichiometry   41
Stone Age   213
Stoney, George Johnstone   57
Strassman, Fritz   160
strong nuclear force   116
sub-shell   122
sugar   199
Sun   303
superconducor   251
supernova   277, 298
synchrotron radiation   265
synchrotron   263
synthetic fibres   238, 240
Szilárd, Leó   163

Taniguchi, Norio   250
tau neutrino   268
tau particle   268
technetium   139
terminal velocity   63
thalidomide   209
The Sceptical Chymist   29
thermistor   247
Thomson, George Paget   101
Thomson, Joseph John   58, 95
thorium   141
Three Mile Island   166
thymine   199
time   281
time-independent Schrödinger wave equation   109
tin   220
Traité Élémentaire de Chimie   39
transmission electron microscopy   175
Trapezium cluster   302
triple bond   48
triple-α process   158, 295
tritium   167
Tube Alloys Project   163
tungsten steel   227
tunnelling   118
turbostratic carbon fibre   243
Tutankhamun   223
type 1a supernova   278

Uhlenbeck, George   131
ultraviolet catastrophe   79
ultraviolet microscope   174
uracil   201
uranium fission   161
uranium hexafluoride   164
uranium isotopes   141
uranium   70, 139
uranium-235   162
urine   49
Uttar Pradesh   224

vacuum pump   27, 53
Vaisheshika   15
valency   47, 125
van den Broek, A.   85
Varley, Cromwell Fleetwood   56
Venus   137
Vikings   226
Villard, Paul   72
virus   186
viscose   239
viscosity   217
visual cortex   169
visual near point   169
vitreous humour   171
von Guericke, Otto   27, 53
von Laue, Max   177
von Lenard, Philipp   80

W and Z bosons   271
Walton, Ernest   259
Watson, James   203
wave equation   108
wave function   109, 110
wave mechanics   111
wave-particle duality   101
weaving   231
Wenzel, Karl   41
white crumb   239

white dwarf   277, 296
Wilkins, Maurice   203
Wilson cloud chamber   145
Wilson, Charles   145
WIMPS (Weakly Interacting Massive Particles)   290
Wind, Cornelius   82
Woman (meteorite)   223
wool   231
work hardening   220
wrought iron   225

X and Y chromosomes   209
X-radiation   88

X-ray crystallography   265
X-ray diffraction   176, 178
X-ray spectrometer   180
X-rays   65

yarn   231
yellow crumb   239
young star   293
Yukawa, Hideki   258, 268

zero-point energy   110
zincblende   178, 180
Zulu   225
Zweig, George   269